NATO ASI Series

Advanced Science Institutes Series

A series presenting the results of activities sponsored by the NATO Science Committee, which aims at the dissemination of advanced scientific and technological knowledge, with a view to strengthening links between scientific communities.

The Series is published by an international board of publishers in conjunction with the NATO Scientific Affairs Division

A Life Sciences B Physics	Plenum Publishing Corporation London and New York
C Mathematical and Physical Sciences D Behavioural and Social Sciences E Applied Sciences	Kluwer Academic Publishers Dordrecht, Boston and London
F Computer and Systems Sciences G Ecological Sciences H Cell Biology I Global Environmental Change	Springer-Verlag Berlin Heidelberg New York London Paris Tokyo Hong Kong Barcelona Budapest

NATO-PCO DATABASE

The electronic index to the NATO ASI Series provides full bibliographical references (with keywords and/or abstracts) to more than 30000 contributions from international scientists published in all sections of the NATO ASI Series. Access to the NATO-PCO DATABASE compiled by the NATO Publication Coordination Office is possible in two ways:

- via online FILE 128 (NATO-PCO DATABASE) hosted by ESRIN, Via Galileo Galilei, I-00044 Frascati, Italy.
- via CD-ROM „NATO Science & Technology Disk" with user-friendly retrieval software in English, French and German (© WTV GmbH and DATAWARE Technologies Inc. 1992).

The CD-ROM can be ordered through any member of the Board of Publishers or through NATO-PCO, Overijse, Belgium.

Series I: Global Environmental Change, Vol. 14

The ASI Series Books Published as a Result of Activities of the Special Programme on Global Environmental Change

This book contains the proceedings of a NATO Advanced Research Workshop held within the activities of the NATO Special Programme on Global Environmental Change, which started in 1991 under the auspices of the NATO Science Committee.

The volumes published as a result of the activities of the Special Programme are:

Vol. 1: **Global Environmental Change.** Edited by R. W. Corell and P. A. Anderson. 1991.
Vol. 2: **The Last Deglaciation: Absolute and Radiocarbon Chronologies.** Edited by E. Bard and W. S. Broecker. 1992.
Vol. 3: **Start of a Glacial.** Edited by G. J. Kukla and E. Went. 1992.
Vol. 4: **Interactions of C, N, P and S Biogeochemical Cycles and Global Change.** Edited by R. Wollast, F. T. Mackenzie and L. Chou. 1993.
Vol. 5: **Energy and Water Cycles in the Climate System.** Edited by E. Raschke and D. Jacob. 1993.
Vol. 6: **Prediction of Interannual Climate Variations.** Edited by J. Shukla. 1993.
Vol. 7: **The Tropospheric Chemistry of Ozone in the Polar Regions.** Edited by H. Niki and K. H. Becker. 1993.
Vol. 8: **The Role of the Stratosphere in Global Change.** Edited by M.-L. Chanin. 1993.
Vol. 9: **High Spectral Resolution Infrared Remote Sensing for Earth's Weather and Climate Studies.** Edited by A. Chedin, M.T. Chahine and N.A. Scott. 1993.
Vol. 10: **Towards a Model of Ocean Biogeochemical Processes.** Edited by G. T. Evans and M. J. R. Fasham. 1993.
Vol. 11: **Modelling Oceanic Climate Interactions.** Edited by J. Willebrand and D. L. T. Anderson. 1993.
Vol. 12: **Ice in the Climate System.** Edited by W. Richard Peltier. 1993.
Vol. 13: **Atmospheric Methane: Sources, Sinks, and Role in Global Change.** Edited by M. A. K. Khalil. 1993.
Vol. 14: **The Role of Regional Organizations in the Context of Climate Change.** Edited by M. H. Glantz. 1993.
Vol. 15: **The Global Carbon Cycle.** Edited by M. Heimann. 1993.
Vol. 16: **Interacting Stresses on Plants in a Changing Climate.** Edited by M. B. Jackson and C. R. Black. 1993.

The Role of Regional Organizations in the Context of Climate Change

Edited by

Michael H. Glantz

Environmental and Societal Impacts Group
National Center for Atmospheric Research
Boulder, CO 80307-3000
USA

Springer-Verlag
Berlin Heidelberg New York London Paris Tokyo
Hong Kong Barcelona Budapest
Published in cooperation with NATO Scientific Affairs Division

Proceedings of the NATO Advanced Research Workshop on the Role of Regional Organizations in the Context of Climate Change, held at Paris, France, October 5-8, 1992

ISBN 3-540-57252-X Springer-Verlag Berlin Heidelberg New York
ISBN 0-387-57252-X Springer-Verlag New York Berlin Heidelberg

Library of Congress Cataloging-in-Publication Data
The Role of regional organizations in context of climate change / edited by Michael H. Glantz. p. cm. – (NATO ASI series. Series I, Global environmental change ; vol. 14)
"Published in cooperation with NATO Scientific Affairs Division."
"Proceedings of the NATO Advanced Research Workshop on the Role of Regional Organizations in the Context of Climate Change, held at Paris, France, October 5-8, 1992" – T.p. verso.
ISBN 0-387-57252-X (New York : alk. paper). – ISBN 3-540-57252-X (Berlin : alk. paper)
1. Climatic changes–Environmental aspects–Congresses. 2. Global warming–Environmental aspects–Congresses. 3. International agencies–Congresses. 4. Water resources development–Congresses. I. Glantz, Micheal H. II. North Atlantic Treaty Organization. Scientific Affairs Division. III. NATO Advanced Research Workshop on the Role of Regional Organizations in the Context of Climate Change (1992 : Paris, France) IV. Series. QC981.8.C5R657 1994 354.1'855516–dc20 93-27582

Printed in Germany

Typesetting: Camera ready by authors
31/3145 - 5 4 3 2 1 0 - Printed on acid-free paper

The Role of Regional Organizations in the Context of Climate Change

NATO Advanced Research Workshop | Hotel Talleyrand, Paris, France | 5–8 October, 1992

The montage on the preceding page is made up of several *Economist* magazine covers. The specific covers used relate to regionalism.

FOREWORD

The past two decades have seen a remarkable broadening of interest in global warming from a research concern on the part of a limited number of scientists to a political problem on a worldwide scale. The nature of this transformation would itself be a fruitful study for a mixed team of social scientists and natural scientists. It would be valuable to assess the differing nature of the staging posts along this road: the First World Climate Conference in 1979, which was a meeting of scientists talking to scientists; the Villach Assessment of 1985, which was a meeting of scientists whose report was given attention by the policy advisers of a number of governments; the Second World Climate Conference of 1990, which consisted of a scientific meeting followed by a Ministerial Meeting; and the United Nations Framework Convention on Climate Change of 1992 signed by 158 countries at the UN Conference on Environment and Development in June 1992.

The present publication is a welcome contribution of the followup to UNCED. By focusing on a specific problem, it avoids the pitfall of undue generalization and provides the basis for fruitful discussion between natural scientists, social scientists, and policymakers. To choose as the area of concentration a particular scale also helped to produce meaningful discussion likely to lead to action. In both the natural and the social sciences, the question of scale is of key importance as different types of phenomena occur at different scales and different explanatory laws are formulated at these separate scales. Political structures and procedures also differ widely at national, regional, and global scales.

The regional scale poses many problems for all the groups interested in global warming. The regional effects of the increase in greenhouse gases are known with less certainty than the global effects. The suitability of the present regional organizations, established and designed on the basis of other economic and political priorities, to deal with problems of environment and development has to be carefully evaluated and some adaptations may be necessary. The present publication contains an important review both of these general problems and of specific problems related to freshwater resources and to marine resources. It provides an excellent starting point for further work on these important topics. The organizers and authors are to be congratulated on what has been achieved.

James C.I. Dooge
President
International Council of Scientific Unions

PREFACE

The focus of the NATO Advanced Research Workshop on "The Role of Regional Organizations in the Context of Climate Change" was to assess the potential role for regional international organizations in the context of global warming. Experts from four areas of interest presented their views on regional organizations, transboundary water resources, transboundary marine fisheries and climate change research. The participants represented expertise drawn from regions and regional organizations around the globe, and from the social and physical sciences. The climate change issue was discussed at some length with the main speaker highlighting the known and uncertain aspects of the issue. Another key speaker presented some of the political aspects of global warming. The regional implications of global climate change were discussed at length, followed by a general overview of regional organizations of interest to the ARW participants.

Uncertainties in the scientific findings related to global warming are not insignificant and require considerably more research to reduce or eliminate them (such as cloud feedback mechanisms; will they reinforce greenhouse gases-induced global warming or mitigate it). It became clear in these discussions that many mid- and high-level government officials in many countries are not yet convinced that global warming is THE most important problem that their countries face today, with the exception of island nations subject to the threat of sea level rise and inundation.

Historical and contemporary transboundary water resource issues were discussed with emphasis, by example, on the Danube, the Euphrates, the Colorado and the Indus rivers. Discussions followed about plausible climate change scenarios in which regional organizations might have a role to play in resolving transboundary resource-related controversies. Features of regional organizations that might make them important to the climate change issue were identified and assessed.

Attention then shifted to climate change and fisheries issues and the role of regional organizations in the management of climate-related transboundary fisheries' conflicts. Participants focused on speculation about the regional changes in fisheries resources that might be expected with or without a change in the climate and conditions under which regional organizations might play major roles in resolving fisheries controversies.

The participants provided numerous examples of how regional organizations have coped (or failed to cope) with environmental change at the regional level. A sobering overview of the limitations of regional organizations in general was presented and discussed. Regional organizations are subject to the political influences of their member states; they are often imbalanced with regard to national influence within the organization. The dominant view was that existing regional organizations should be first considered for potential roles in the context of global warming, rather than creating new ones. Given that regional organizations (political, resources-related, economic, geographic, or functional) have considerable experience in dealing with a variety of issues (if not climate change per se), they could expand their activities relatively easily to include regional issues related to climate change. For example, all countries are concerned with economic development planning. Climate change scenarios derived by a variety of methods (general circulation models, historical analogs, and paleoclimatic reconstructions) could serve to enhance the development planning process.

The changing world order brought about by the collapse of communism in Eastern Europe and the Soviet Union, coupled with the rising concern for the global environment in general and global warming in particular, has provided a window of opportunity for regional organizations to reassess their traditional functions in order to expand them to incorporate activities related to some aspects of the climate change issue (ranging from the forecast of regional climate change to assessing its impacts to developing responses).

Acknowledgments

I would like to express my gratitude to Jan Stewart, for taking care of the logistics of the Paris workshop and in her shepherding of the manuscripts from draft to final production. I would also like to thank our Paris hosts, the US Information Service of the US Embassy, in particular Frank Huffman and Sylvie Vacheret, who gave us access to their facilities and help in ways too numerous to mention. Martin Price, who took on the onerous task of Rapporteur for the workshop, deserves special thanks. Isabel Dolan was tireless as both a member of the Secretariat and as the person responsible for arrangements in Paris. I would also like to thank Brad Abrahamson, research assistant. Finally, I would like to express my appreciation to Dr. Luis da Cunha, Scientific and Environmental Affairs Division of NATO, for his support of our Advanced Research Workshop.

Michael H. Glantz

CONTENTS

Foreword VII
Preface IX
Glossary of Acronyms 199
Introduction 1

PART I: **Executive Summary** 7

PART II: **Summary of Discussion Sessions**

Michael H. Glantz and Martin F. Price
Science Issues 33
Policy Issues 36
Societal Aspects 41
Regional Organizations 45
Non-Governmental Organizations 53

PART III: **Discussion Papers**

Background on the Climate Change Issue
Michael H. Glantz
The regionalization of climate-related environmental problems 57
Jag S. Maini
The role of regional organizations in addressing climate change and other complex environment and development issues 62
Robert S. Kandel
The climate change issue: Scientific aspects 65
James P. Bruce
The climate change issue: Policy aspects 68
Lynton K. Caldwell
Transnational regional responses to global climate change: Options, obstacles, opportunities 73
Renat Perelet
An environmental security dimension of global climate change 80

Water Resources
Luis V. da Cunha
Climate change and water resources 88
Ödön Starosolszky
Regional organizations and climate-related changes in the water regime 93
James L. Wescoat, Jr.
Climate change and international water problems: Issues related to the formation and transformation of regional organizations 96

Ilter Turan
Regional organization for water utilization in the Middle East 103
Rui Roda
Transboundary water resources on the Iberian Peninsula 109

Marine Resources
Gunnar Kullenberg
Some aspects of regional cooperation in the marine sciences 113
Edward Miles
The potential role of regional organizations related to the marine environment in the context of global climate change 120
Jean-Paul Troadec
Scarcity, property allocation, and climate change 128

Regional Organizations
Newton V. Cordeiro
The Organization of American States (OAS) and issues of environment and development 136
Sam Tewungwa
Regional organizations and environmental change: An East African example 141
Saleemul Huq
The South Asian Association for Regional Cooperation (SAARC) and climate change 149
Arild Underdal
The roles of IGOs in international environmental management: Arena or actor? 153
Naresh C. Singh
Role of regional organizations in the context of climate change 159

PART IV: **Appendix**

Biographical Statements of Participants 169
Item 4, Provisional Agenda, Working Group III 185
List of Participants 191
Suggested Bibliography 193
Glossary of Acronyms 199
Subject Index 203

Introduction

In addition to NATO's traditional focus on political and military affairs, activities carried out by its Scientific and Environmental Affairs Division around a third dimension which encompasses environmental and societal issues. Possible areas for action through its Science Committee include, but are not limited to, the following: 1) bridge-building among disciplines; 2) consideration of a broader range of topics for research and meetings; 3) the involvement of more people and the request for more proposals from researchers in non-NATO countries; 4) bringing together representatives of regional organizations from within and outside the NATO area. Any one of these changes could help to reduce the potential for regional conflict, which is one of NATO's primary goals.

In the spirit of these broad aims, a workshop was convened on the potential roles of regional organizations in the context of climate change in October 1992. The workshop sought to identify key issues and problems to be addressed by regional organizations, with special emphasis on water resources and fisheries. Because the workshop was issues-oriented, no attempt was made in the papers presented, or in this volume, to review or cite the voluminous scientific literatures in the fields of global change, regional organizations, international resources, and fisheries. Instead, participants focused on key questions that need to be addressed, and lessons that might be shared among the fields of expertise involved.

The workshop was opened by the US Ambassador to France, Walter J.P. Curley, who welcomed the participants to Paris and to the US Embassy for this Advanced Research Workshop.

Ambassador Curley remarked that environmental issues have become increasingly important in recent years, as people and their governments come to understand the interrelationships between the environment, on the one hand, and economics, society, and politics on the other. More specifically, he then noted that the causes and potential effects of global climate change have been much debated in scientific circles in recent years, and he was particularly pleased that the workshop was conducted in Paris. He alluded to this conference, organized by the prestigious National Center for Atmospheric Research, supported by a grant from NATO, and hosted by the US Information Service at the Paris Embassy, as an example of the leadership of the United States in pursuing solutions to the problems of the environment, not only domestically but also in the global context. He wished the participants the best of luck in their deliberations.

Luis da Cunha, Science Administrator, NATO, then welcomed the participants on behalf of NATO's Division of Scientific and Environmental Affairs, which sponsored the Advanced Research Workshop (ARW) through the NATO Science Committee's

Special Program on the Science of Global Environmental Change. He noted that two of the principal goals of this program were to advance science and technology in NATO countries, and to strengthen the links among scientists on both sides of the Atlantic Ocean as well as between NATO and central and eastern European countries. He highlighted some key components of the program which include fellowships, grants, meetings, and targeted programs. Da Cunha noted that about 12,000 scientists become involved in the activities of NATO's science program each year. This particular workshop was highlighted as being unusual among Advanced Research Workshops because of wide-ranging expertise present, drawn not only from various natural and social science and academic circles, but from the policymaking community as well.

Following da Cunha's opening remarks, Michael Glantz, Program Director of the Environmental and Societal Impacts Group (ESIG) at the National Center for Atmospheric Research (NCAR) and workshop organizer, presented the schedule of the workshop. He suggested that the rapid economic, social, and political changes that accompanied the end of the Cold War and the increasing political concerns about global environmental change might provide regional organizations with a significant "window of opportunity" to review their traditional functions and activities with a view either to initiate or enhance, as the case may be, their roles in issues related to the regional implications of global environmental change. Glantz noted that the workshop participants represent a broad range of expertise, particularly in the fields of climate impacts, water resources, fisheries, and regional organizations. Each participant was then asked to briefly share his expertise and experiences to provide some views on the best role(s) – if any – for existing or yet-to-be-established regional organizations with respect to environmental issues, particularly global warming.

Participants

Michael H. Glantz .... Conference Organizer; Environmental and Societal Impacts Group (ESIG), National Center for Atmospheric Research, USA

Martin F. Price .... (Rapporteur) Environmental Change Unit, University of Oxford, Oxford, UK

James P. Bruce .... Canadian Climate Centre, Ontario, Canada

Lynton K. Caldwell ...... University of Indiana, Bloomington, Indiana

Newton V. Cordeiro ..... Organization of American States, Washington, DC

Luis V. da Cunha .... Scientific & Environmental Affairs Division, NATO, Brussels, Belgium

Saleemul Huq .... Bangladesh Centre for Advanced Studies, Dhaka, Bangladesh

Robert S. Kandel .... Laboratoire de Meteorologie Dynamique, Ecole Polytechnique, Paris, France

Gunnar Kullenberg ...... Intergovernmental Oceanographic Commission, Paris, France

Jean Labrousse .... Terre, Ocean, Espace et Environnement, Paris, France

Jag S. Maini .... Forest Environment, Forestry Canada, Ottawa, Canada

Edward Miles .... School of Marine Affairs, University of Washington, Seattle, Washington

Renat Perelet .... Institute for Systems Analysis, Russian Academy of Sciences, Moscow, Russia

Natalie Philippon .... Intergovernmental Oceanographic Commission, Paris, France

Rui J.M.R. Roda .... Empresa Portuguesa das Aguas Libres, Lisbon, Portugal

Naresh C. Singh .... Caribbean Environmental Health Institute, Castries, St. Lucia

Ödön Starosolszky .... Water Resources Research Center (VITUKI), Budapest, Hungary

Sam Tewungwa .... Intergovernmental Panel on Climate Change (IPCC) Secretariat, World Meteorological Organization, Geneva, Switzerland

Jean-Paul Troadec .... Centre ORSTOM, Plouzane, France

Ilter Turan .... Istanbul University, Istanbul, Turkey

Arild Underdal .... Center for International Climate & Energy Research, Oslo, Norway

James L. Wescoat, Jr. .... University of Colorado, Boulder, Colorado

Secretariat:

Frank Huffman .... US Information Service, Paris, France

Sylvie Vacheret .... US Information Service, Paris, France

D. Jan Stewart .... Secretary, ESIG, National Center for Atmospheric Research

Isabel Dolan .... Paris liaison, France

Bradley Abrahamson ... ESIG research assistant, University of Colorado, Boulder, Colorado

The workshop was divided into sessions, each of which focused on a different issue. The sessions, in order of occurrence, were as follows:

Monday, 5 October

Session 1: The climate change issue: The science

Session 2: The climate change issue: The policy

Session 3: Regional implications of climate change

Session 4: Introduction to regional organizations

Tuesday, 6 October

Session 5: Contemporary water resource-related problems

Session 6: The role of regional organizations in the management of contemporary water issues

Session 7: Changes in the distribution of regional water resources that might be expected in the future (even in the absence of a climate change)

Session 8: Water resources and climate change

Session 9: Plausible scenarios for which regional organizations could play a major role in resolving water resource controversies

Session 10: Features of regional organizations that make them important to the climate change issue

Wednesday, 7 October

Session 11: The role of regional organizations in the management of fisheries-related conflicts today

Session 12: Changes in fisheries resources that might be expected in the future (even in the absence of a climate change)

Session 13: Plausible scenarios for which regional organizations could play a major role in resolving fisheries controversies

Session 14: What role for new (or existing) regional organizations in the context of climate change

Session 15: Examples of regional organizations that have coped with environment-related change

Session 16: Examples of regional organizations that could not cope with climate-related environmental change

Thursday, 8 October

Session 17: UN agencies, regional organizations, and environmental issues

Session 18: Is there an ideal regional mechanism that could be devised to prepare for coping with the impacts of changes in resources resulting from a climate change?

Session 19: Workshop recommendations and publication prospects

* * *

Recognizing the existing extensive literatures on each of the various aspects of interest to the ARW, each participant was asked to prepare a short background paper, providing some views on regional organizations, fisheries, water resources and/or global warming. These papers were made available to participants in advance of the meeting. During the ARW, participants were asked to make opening comments for the purpose of stimulating discussion for that session.

Part I is an Executive Summary of the discussion papers prepared by the workshop participants, along with key points of those papers. The comments made in the discussion sessions which followed each presentation were combined into one section, Part II, and clustered by topical theme according to the perceptions of the ARW organizer – science issues, policy issues, societal aspects, regional organizations, and nongovernmental organizations. Several of the items placed in each of these categories could have appeared elsewhere; these are not mutually exclusive categories.

In the next section of the workshop report, Part III, the opening comments of the different participants have been merged into their discussion papers. Part IV contains brief biographical statements about the participants, the ARW agenda, Agenda 21's Item 2(c) of Working Group III, a suggested bibliography, and a subject index.

PART I

Executive Summary

The Regionalization of Climate-Related Environmental Problems

Michael H. Glantz

It appears that the euphoria generated by the end of the Cold War may have been somewhat premature, given the numerous conflicts presently under way in all parts of the world, even in Europe. Perhaps the 1990s provide a "window of opportunity" for an objective assessment of regional organizations and their potential contribution toward resource-related cooperation and conflict resolution. Given the resurgence of worldwide concern about "global" environmental issues and their regional causes or consequences, such organizations could provide an effective arena for discussing, resolving or averting regional conflicts related to environmental change. Whereas regional international organizations (functional as well as geographic) have often been relegated to marginal roles in the international political arena with regard to resources issues, this will likely change in the future. Such a change can be expected because, as climate regimes shift in response to global warming, so too will the location of some highly valued resources, sometimes across national boundaries. As resources and people dependent on them "migrate" on land and in the marine environment, the risks of regional conflicts, as well as the opportunities for regional cooperation, are likely to increase. The time may be right to talk about the "regionalization of environmental problems."

Key Points

- Climate modelers are not yet in a position to provide societal and environmental impacts researchers with the regional detail needed for the construction of reliable scenarios. Thus, national and subnational policymakers have become increasingly concerned about how a global climate change might manifest itself at the regional and local levels.

- Questions about the potential severity of impacts of global climate change at the regional and local levels pertain not only to climatic factors (including extreme events), but also to their interactions with societies possessing different capabilities for implementing proactive as well as reactive policy responses.

- Given that the frequency of such conflicts (or opportunities for cooperation) could increase in the face of global climate change and its regional consequences (e.g., the shifting of highly valued resources), increased attention should be given to the role of regional organizations (both existing as well as those that might yet be established) in resource-related regional conflict resolution.

The Role of Regional Organizations in Addressing Climate Change and Other Complex Environment and Development Issues

Jag S. Maini

Climate change is only one of several large-scale environment and development issues that has recently appeared on international scientific, policy, and political agendas. Attempts at addressing these issues require, first, the formulation of a common understanding of the issues and, second, a commitment to work collectively at the national, regional, and international levels.

Addressing global-scale issues such as climate change has regional and national dimensions, as well as global ones, in terms of causal factors, consequences, and corrective (i.e., preventive, mitigative, or adaptive) measures. For example, more rapid progress may be made at the regional level to address the issue of the health of the oceans and seas. Similarly, the issue of the conservation and sustainable development of global forests may be addressed on a "regional" basis, such as the Amazon Basin, the humid tropics in western Africa, and the circumpolar boreal forests zone.

Key Points

- How a regional organization approaches an issue is determined largely by its mandate and its policy framework.
- In order to avoid costly mistakes and delays in the implementation of corrective actions, the scientific and policymaking communities must gain insights from past experiences in coping with other situations of change.
- The mobilization of regional organizations may prove to be a critical factor in addressing complex issues related to environment and development such as global climate change.

The Climate Change Issue: Scientific Aspects

Robert S. Kandel

Climate on the regional scale is governed by various aspects of the global climate system which is constantly changing; the critical issue concerns the rapidity with which ecologically significant global and regional changes are to be expected.

There is no doubt that the continuing increase in greenhouse gases must enhance trapping of infrared radiation in the atmosphere and lead to global warming, but how much, and when? The extreme uncertainty of model estimates of regional climate changes expected to accompany a global warming means that it will be difficult to identify such a trend with confidence, as long as temperature and precipitation changes are small. In other words, the CO_2 "fingerprint" that scientists and policymakers hope to identify remains "blurred." If there is no convincing progress in model estimates, it will be difficult to estimate climate sensitivity before changes become significant. By then, it will probably be harder to reduce greenhouse gas emissions at a pace fast enough to avoid still larger, unwanted climate changes. The uncertainty that surrounds model estimates also means that, in seeking to develop societal adaptation strategies to, rather than prevention of, climate change, planners can only consider the regional climate change estimates as plausible but not necessarily reliable scenarios of future change. Any particular scenario may have quite different implications for different regions; and different plausible scenarios may have very different implications for a given region.

Key Points

- Examination of the ecological and societal impacts of climate change must focus on the effects of climate changes at the regional level.
- The activities which produce greenhouse gases differ greatly from one region to the next. However, because of the rapidity of atmospheric circulation and the relatively long residence times of these gases in the atmosphere, greenhouse gas enrichment is global.
- Despite reports on consensus, there remains considerable disagreement on such important issues as (i) the likely course of changes in greenhouse gas concentrations in the next several decades, and how different policies might affect it; and (ii) the sensitivity of the climate system, in particular as moderated or amplified by water vapor and cloud feedbacks, to changes in greenhouse gas abundances.
- Although modern climate models are more or less successful in simulating observed features of the geographical distribution of regional climates and their seasonal cycles, they disagree in many aspects of their predictions of regional climate changes, and remain inadequate in their treatment of ocean–atmosphere coupling and of the water cycle. As a result, little confidence can yet be given to model projections of regional climate changes that might accompany global warming.

The Climate Change Issue: Policy Aspects

James P. Bruce

According to the Intergovernmental Panel on Climate Change (IPCC's) "business as usual" scenario, rates of CO_2 emissions from developing countries are expected to continue to increase faster than those from industrialized countries, although in total the industrial economies will continue to emit far more greenhouse gases. This implies that, if the atmospheric concentration of CO_2 is to be stabilized at some point in the future, in addition to industrialized countries decreasing their dependence on fossil fuels, future paths of development (involving technology transfer) in other countries should be designed to minimize CO_2 and other greenhouse gas emissions (GHGs).

Regional organizations can play a key role in exchange of national views and experiences on emission limitation options. Regional organizations come in a bewildering array of sizes and functions, and their roles in the climate change issue can be equally diverse. It may be useful to consider their roles within the framework of the following activities associated with climate change: data collection, climate system research, impact assessments, adaptation strategies, and control programs. To these ends, a number of regional organizations have already proven effective in facilitating exchanges of information and perspectives.

Key Points

- Because some degree of global warming can be expected to occur due to past and continuing human activities, adaptive measures designed to cope with regional changes of global warming are essential for climate-sensitive sectors such as agriculture, water resources,and natural hazards.

- In some areas a regional approach can be helpful in devising appropriate response strategies. This is especially true, for example, in connection with resolution of water resource conflicts, sea level rise, and reduction of losses due to natural disasters.

- Because of the intense economic and political sensitivity and the disparities among countries in their sources of emissions of GHGs, countries within a region can learn from each other about the benefits, costs and feasibility of various emission reduction and action strategies.

Transnational Regional Responses to Global Climate Change: Options – Obstacles – Opportunities

Lynton K. Caldwell

Regional organizations are created by states because they find it increasingly difficult to deal effectively with numerous international issues of varying levels of national importance that require constant attention. Regional organizations have four advantages: flexibility of collective action; capability to initiate policy; negotiation; and exchange of views.

Although rapid by geological standards, in human terms global climate change as presently projected appears to be slow and, for the present, unobtrusive, since noticeable effects are not expected to occur for several more decades. It is not in the nature of practical politics to give much attention to circumstances that are not presently compelling or that are not likely to arouse public concern in the near future. In a world of many troubles, politicians are reluctant to ask for present-day sacrifices on behalf of a future (an uncertain one) that most of their constituents will not live to see. However, the technical and behavioral changes that could arrest global warming are known, and many would merit adoption even if global warming were not an issue.

A rational obstacle to national responses to transnational problems is an honest uncertainty as to the most feasible and effective way to address or resolve those problems. The obvious inclination of most politicians and public officials is to adapt existing national institutions to the problem. When this option seems insufficient, the usual inclination is to proceed with caution, thus creating transboundary or regional bodies with strictly delimited powers and provisions for national noncompliance on disputed propositions. The emergence of an international environmental movement and network of non-governmental organizations (NGOs) has begun to remove the obstacle of an insufficiently organized transnational constituency able to command the attention of governments.

Key Points

- We may identify options for new transboundary arrangements and the opportunities that alternative choices present. But a realistic approach to these conjectures requires the identification of the obstacles to regional and national cooperation and the devising of credible alternatives to institutional inadequacies.

- The historical record demonstrates that the NEED for basic changes in social institutions characteristically emerges well in advance of political recognition of the NECESSITY for such change.

- The reluctance to "borrow trouble" from the future is one aspect of a larger phenomenon – "institutional lag." The lag syndrome is the sum of a number of evasive responses to threatening developments in society that might imply a need for new institutional arrangements.

- National agencies are seldom ready to subordinate their autonomy to transnational superordinate authority. In nearly every case, goal attainment by regional organizations is thwarted or delayed because of "foot-dragging" tactics by governments and their "turf conscious" agencies.

- The emergence of an international environmental movement and network of NGOs has created a transnational constituency able to command the attention of governments. It was hardly a sudden conversion to "environmentalism" that brought more than one hundred heads of state to the UN Conference on Environment and Development.

An Environmental Security Dimension of Global Climate Change

Renat Perelet

The uneven local and regional impacts of global environmental changes such as climate variability and the prospects of a global climate change have increasingly contributed to tension within states and between them and present a new challenge to regional organizations. NATO's new strategic concept laid out in the Rome Declaration (1991) stressed that security risk comes significantly nowadays not from premeditated aggression but from negative consequences of various difficulties, including those relating to their environmental component.

Environmental issues are wittingly tied into other problems and often used as bargaining chips in international negotiations. Regional environmental threats resulting from climate change may affect traditionally "free" natural assets (e.g., soil fertility and, therefore, food production), as well as agricultural investment and human settlement patterns. Global environmental changes are essentially of a transboundary nature, affecting the well-being (in fact, security) of nation-states. The human responses to them (e.g., an increase in environmental refugees, changing fishery practices) spill over national frontiers and can effectively be dealt with internationally, especially through actions by regional organizations coordinated globally.

National environmental security (and sovereignty) can be affected by at least three major environmental factors: 1) transboundary, regional, or global environmental change; 2) international trade relationships; and 3) environmentally unsound development strategies. The concept of security has evolved beyond military and political security to include the environmental component. The well-being of a nation-state (as long as the latter exists) and the international community depends not only on political, military and economic capacity, but increasingly on environmental security. The possible effects of climate change have evolved from scientific to economic and political to, most recently, security issues.

Key Points

- The environment has become a legitimate national and international security issue. Thus, a new definition of security should include, in particular, an appreciation of the importance of climate-change-related scarcity of natural assets and adverse effects of attendant environmental changes on human health and well-being. The threats of both environmental resource scarcity and change can lead to international tension and conflict. They call for achievement of environmental resource (eco)sufficiency and (eco)parity in terms of just access to environmental assets (e.g., the buffering of international freshwater reservoirs, food reserves, etc.), resources and the environment as key components of national or regional security.

- Four major factors that affect international security in the case of global environmental change (GEC) and need to be negotiated among states and regions stem from: differing perceptions of GEC by government leaders, the uneven impact of GEC on countries, heterogeneous contributions of countries to GEC, and a disparity between national or regional capacities to respond to adverse GEC.

- Some ostensibly domestic problems (some of which are likely to be related to climate change) become global policy issues, when they affect ecosystems of international significance, and other countries take a special interest in them (e.g., tropical deforestation in the Amazon, pollution of Lake Baikal, the drying out of the Aral Sea). They cause international concern that may lead to tension.

Climate Change and Water Resources

Luis V. da Cunha

According to atmospheric general circulation models, a doubling of CO_2 in the atmosphere could result in an increase in global average temperature of about 3°C, bringing about changes in precipitation and evaporation and, thus, also in runoff. The magnitude of such changes in water resources as well as changes in its distribution are expected to have significant social, economic and environmental impacts. Researchers need to identify possible difficulties water managers might face, and to identify ways that current water resources management and planning strategies might need to be modified in order to mitigate, if not prevent, the adverse impacts or to take advantage of favorable ones that might be associated with a climate change.

Case studies can be an instructive approach to gaining a glimpse of the future. Although no two regional situations are alike, each individual study can identify potential problems that might emerge in analogous situations. Each study can help to understand processes related to the allocation of often scarce water resources, as well as in uncovering perceptions about climate variability and change and how those perceptions might affect decisionmaking processes regarding transboundary water resources.

Key Points

- It is important and useful to address the potential magnitude of climatic change impacts on regional water resources. This will enable researchers to identify possible difficulties water managers might face, and to identify ways that current water resources management and planning strategies might be modified.

- The impacts of a climate change on water resources will not only depend on changes in climatic mean values, but also on changes in climatic variability. The aspects of climate variability that tend to attract the attention of policymakers and the public are changes in the frequency, intensity, and location of extreme events such as droughts and floods.

- The most critical effect of regional climate change is likely to be on the timing and magnitude of runoff and fluctuations in soil moisture at the regional level. As yet, the uncertainty that accompanies GCM-based water resources scenarios makes them difficult to use in the development of realistic management strategies.

- There is concern about the possibility of an increase in "water wars" and perhaps a growing need for regional organizations to cope with such potential transboundary resource-related conflict and cooperation.

Regional Organizations and Climate-Related Changes in the Water Regime

Ödön Starosolszky

Hydrologists have only recently become interested in global warming and its possible impacts on environment and society. The projected changes in extremes such as maximum and minimum flows will have significant implications for the design of hydraulic structures, particularly because runoff is a nonlinear process. If regional organizations are to respond to the potential consequences of global warming, they will need to increase their effectiveness. The development of new regional organizations may be necessary. The experience of regional organizations in the Danube River basin might serve as a relevant example in designing new regional organizations or restructuring existing ones. Some climatologists have predicted that climate change could result in an increase in the Danube Basin in the frequency and intensity of extreme climate events and consequently in related extreme hydrologic events.

Regional organizations should undertake pilot projects on selected river basins (e.g., over the Danube and Rhine, within Europe) to demonstrate the potential role of regional organizations in the mitigation of harmful effects that would likely accompany changes in the regional water balance that might result from global warming.

Key Points

- Donor organizations should consider more seriously how to assist regions to mitigate the adverse impacts on water resources of climate change (e.g., by constructing dams, diversions, reservoirs, etc., when deemed appropriate and necessary).

- More attention should be devoted to the regional changes that may adversely affect those regions already considered to be extremely vulnerable (e.g., Tigris-Euphrates, Ganges-Brahmaputra, Nile, Niger, La Plata), where changes in water regimes may generate political or economic conflicts among riparian countries.

- Assumptions about climate must be made more specific in discussions about hydrological property rights, or each side in a dispute will tend to adopt the scenario that is best for negotiation from its point of view.

Climate Change and International Water Problems: Issues Related to the Formation and Transformation of Regional Organizations

James L. Wescoat, Jr.

Research into the implications of climate change for international water resource management, as an emerging field of study, has given little attention to the role of regional intergovernmental organizations in addressing, resolving, or exacerbating water problems. The history of modern regional organizations in international water resources management extends back in time at least two centuries. A brief examination of the evolution of such organizations could provide some insights, analogies and perhaps lessons for regional organizations in the future.

Regional organizations are seen by the international policymaking community as important, but their actions are highly limited by the management of water resources at national and subnational levels.

Key Points

- Recent global environmental policy initiatives (including UNCED, the International Conference on Water and Environment, and the International Law Commission) have given surprisingly little attention to the role of regional organizations in international water management.

- One plausible explanation for the current neglect of regional organizations in international water policy arenas is that new regional organizations tend to be formed during periods of territorial *integration* (e.g., in the aftermath of war or the formation of political and economic alliances). The post-Cold War period may eventually lead to territorial (and thereby water resources) integration in some regions, but not in regionals of territorial fragmentation.

- Recent research on "forecasting by analogy" has contributed to the construction of climate change scenarios, assessment of potential impacts, and identification of adjustments that might be overlooked or oversimplified by other methods. We need to focus in a clear way on some of the following major types of analogy: historical, inter-regional, inter-sectoral, and complex combination of these analogies.

Regional Organization for Water Utilization in the Middle East

Ilter Turan

Climatic change is not the only source of change in the distribution of water resources at the regional scale. Most countries in the Middle East regard water as a resource in which they must be self-sufficient. Water availability is a priority of the highest order, and having to rely on other countries is generally viewed as an unacceptable option. A typical approach is to ensure as large a national supply of water as possible, and then decide how to use it. For these reasons, regional organizations for water resources management are lacking in this particular region, and there is great resistance to change.

The difficulties encountered in getting states to cooperate on questions of transboundary water utilization should not allow us to ignore the need for such cooperation. In searching for ways of how regional cooperation may be enhanced through regional organizations, it may be useful to begin by identifying the areas of possible cooperation.

There is no standard organizational type which is suitable for dealing with all aspects of water, environment and climatic change issues at the regional level. The choice of an inappropriate type of organization may impede rather than enhance the implementation of solutions.

Key Points

- Owing to the speed of change, international law sufficient to provide reasonably clear guidelines in the utilization of transboundary waters has not yet evolved. Similarly, there has been a paucity of international organizations of a global or regional nature whose function it is to study, plan, regulate and administer and even adjudicate the utilization of transboundary waters.

- It seems particularly difficult to get independent states to cooperate on the use of transboundary waters despite the need for such cooperation, because countries display an unusual sensitivity to being in full control of their water resources.

- It may be useful to first identify those areas in which regional cooperation through regional organizations is to take place, such as data collection, water conservation, and the more efficient use of water.

- A regional organization should not be viewed as a panacea which would bring solutions to regional water problems and the problems of environmental and, possibly, climatic change.

Transboundary Water Resources on the Iberian Peninsula

Rui Roda

More than 40% of Portugal's total surface water supply originates in Spain, mainly through the Douro, Tagus and Guadiana rivers. Its availability depends on the exploitation of the existing Spanish reservoirs.

As a downstream user, Portugal has been concerned about the increase in the demands of Spanish upstream water users and the related threat to the quality of its transboundary water supply. Scenarios about possible impacts of climate change in southern Europe suggest that precipitation will decrease in all seasons except autumn. This potential decrease in Iberian precipitation in the spring and summer seasons would generate a major crisis in Portugal, considering present-day chronic water shortages.

Key Points

- The hydrologic year which just ended (Fall 1992) provides insights about socioeconomic difficulties which could occur in the future. It also reinforces the urgent need for joint action between Portugal and Spain.

- The resolution of future climate-change-related problems cannot be based solely on solutions designed to increase water availability. Spain and Portugal must also act jointly to increase the efficient and equitable use of Iberian water resources.

Some Aspects of Regional Cooperation in Marine Sciences

Gunnar Kullenberg

One example of long-standing regional cooperation in marine research and systematic observations is the International Council for the Exploration of the Sea (ICES) initiated at the end of the 1800s. More recently, the preparations for UNCED 92 (Earth Summit) emphasized the importance of the regional scale in the context of Agenda 21 for the oceans. Regional cooperation and coordination is considered vital for the protection and sustainable use of the marine environment and its resources.

The basic ideas of the Intergovernmental Oceanographic Commission (IOC) are that regional cooperation is a must, because of the dynamics of the ocean and the marine environment (as well as its resources); because a regional perspective helps define priorities and priority needs; because of the pooling of efforts; because it fosters regional and national development; and because of the need to address many marine problems on a regional basis in view of the dynamical properties of the ocean environment and its interactions with land and atmosphere. Thus, the "regional window" is natural, as far as marine sciences are concerned.

Key Points

- Use should be made of a regional mechanism to generate support from several donors, on a bilateral or multilateral basis, toward the same regional objective. The bilateral and multilateral aid mechanisms must be seen as complementary and not conflicting.

- An important aspect of future directions in marine science must encompass the integration of open ocean research programs with those on continental shelf seas and in the coastal zones. Consideration of open ocean processes in isolation or separately from the processes occurring in coastal zones will be counterproductive in the face of global climatic changes and sea level rise. This needs to be taken into account in regional cooperative programs.

- Another remedy is that functioning national liaison mechanisms be established which are linked to all relevant ministries and governmental departments. These bodies must be active on the national scene and fulfill the coordination role at the national level.

The Potential Role of Regional Organizations Related to the Marine Environment in the Context of Global Climate Change

Edward Miles

While global climate change is a new issue on the international agenda, it is not a new type of issue (or problem). It falls into the sub-class of global environmental problems of a long time scale where uncertainty is high, where scientific consensus about causes and consequences evolves slowly, where trade-offs are potentially painful in a socioeconomic sense, and where political inertia is great.

The potential role of regional organizations in the context of global climate change is viewed in terms of the experiences to date in responding to other global environmental problems and in seeking to facilitate international collaboration in scientific research, in exploiting while protecting the global commons, and in R&D collaboration in advanced technology. All of these activities are mixed motive "games" (i.e., combining conflict and cooperation) and, therefore, fall within the class of collective action or public goods problems where a well-developed theory already exists.

Scientific questions become high priority under one or a combination of three conditions: a) when there are significant material gains to be had; b) when there are politically significant "bads" to be avoided; or c) when major domestic constituencies compel governments to face up to the issues, even though the leadership and certain interest groups may not wish to do so.

Key Points

- For states, international collaboration is never an end in itself. Collaboration is always purposive, the aim being to solve some problems or realize some gains which states could likely not have achieved, left to their own devices.
- Even when regional cooperation "works," one should expect an inevitable rivalry between the regional level and the bilateral level, because the costs of interdependence of the latter are always less.
- The exclusive economic zone (EEZ) regime limits the role of regional fisheries commissions, so that most have lost their former management responsibilities to the coastal states. The commissions can only be effective when their activities are harmonized with the objectives and policies of specific states.
- With respect to climate change, the following four lessons can be learned from past experience with fisheries institutions: the nature of the decision system is critical; the divisibility of payoffs is critical; regulation requires monitoring and enforcement; and the role of science is crucial.

Scarcity, Property Allocation, and Climate Change

Jean-Paul Troadec

To cope effectively and efficiently with changes in resource abundance and distribution, instruments to adjust fishing rate and capacity to stock productivity are needed. As evidenced by the situation of overcapacity, overfishing, ineffective enforcement of regulations and conflicts among users within and between uses prevailing in a majority of domestic and international fisheries and uses of aquatic ecosystems, those instruments do not match the new situation of resource scarcity. Schematically, such adjustments are effected through three basic mechanisms: (i) social structure of producers, such as in traditional communities; (ii) application of norms and controls by central governments and administrations; and (iii) market mechanisms. When allocation decisions must be taken to share scarce resources, those mechanisms require clear property regimes in order to function efficiently. Recent changes in the ocean regime have been an initial, but partial, move in that direction.

The problem is not one of resource scarcity, but of institutions that are inadequate to cope with the new situation of resource scarcity. The solution to this issue has two facets: i) a physical one: owing to the geographical dimensions of fish stocks; and ii) an institutional one: property regimes and allocation mechanisms need to be clarified.

Climate change will have a number of effects on fisheries. Changes in the ocean structure and circulation will modify the distribution and abundance of individual fish stocks. However, the directions and magnitudes of such changes are not yet known. Improvements of regional circulation models and new research on the relationships between ocean structures and individual fish stock abundance are needed. However, the issue of adjusting the institutions to the new conditions will remain. Institutional innovations and policy formulation are separate issues. Prior adoption of adequate institutions conditions the implementation of important political options. On the other hand, with appropriate institutions, a significant share of the management task could be decentralized, thus lessening the need for central decisions.

Key Points

- In many fisheries, human inputs (capital and manpower) are allocated through economic mechanisms, while access rights are regulated through administrative decisions. When allocation decisions must be taken to share scarce resources, those mechanisms require clear property regimes in order to function efficiently. Thus, the problem with fisheries is not so much one of resource scarcity, but of institutions that are inadequate to cope with the new situation of resource scarcity.

- Currently, the insufficient clarification of property regimes for stocks distributed outside areas of national jurisdiction, and of mechanisms used for allocating use rights to domestic fishermen, are critical weaknesses for adjusting to future changes in fish stock abundance and distribution.

- Examples already exist of a variety of the potential impacts of climate change on fisheries. They provide analogies, when trying to determine how well present fisheries institutions and bodies might cope with future problems.

- By enhancing the mobility of fishermen and vessels with respect to particular fish stocks, the capacity of fleets to adjust to climatic change depends critically on the adoption of institutional innovations.

The Organization of American States (OAS) and Issues of Environment and Development

Newton V. Cordeiro

Concerns over environment and development have always been present in the philosophy and programs of the OAS. Since its creation, the OAS has understood development as an integrated process where physical, biological, cultural, social, economic and political variables continually interact. A number of articles in the OAS Charter address the issues of development and environment and the institutional and operational structure of the organization reflects these concepts. The OAS, as a regional political organization where all 35 independent countries of the hemisphere are members, possesses a comparative advantage in that it brings together both developed and developing countries and provides a regional forum where nations of the north and south have agreed to discuss and seek solutions to the issues of environment and development. Perhaps this political reality reflects the physical and human characteristics of the hemisphere: diversity can lead to complementarity and common solutions.

Key Points

- The OAS is unique among regional organizations because it is the only one that covers an entire hemisphere, from the Arctic to the Antarctic. Within the hemisphere, South America has the largest volume of freshwater flow, amounting to twice the runoff of all the other continents combined.

- OAS member states include some of the world's major producers of carbon dioxide, in both absolute and per capita terms. Also, the Amazon rainforest plays a major role in the global climate system. The Americas include regions which are likely to be among the most as well as the least affected by global warming.

- The OAS possesses a comparative advantage in that it brings together 35 developed and developing countries and provides a regional forum where nations of the north and south to seek solutions to the issues of environment and development.

Regional Organizations and Environmental Change: An East African Example

Sam Tewungwa

Society is now contemplating measures to reduce the pace of the evolution of such environmental problems as acid rain, ozone layer depletion, and global warming of the atmosphere, as well as develop measures designed to prevent, mitigate, or adapt to the changes even before their adverse impacts become manifest. Measures to slow down global warming, for example, are not without costs, which vary from one country to the next. Implementing such measures calls for the pooling together of the world's financial and human resources.

Regional cooperation implies surrender of a certain amount of sovereignty, as collective decisions taken for the good of the whole may not be in the best interests of all of the members all of the time. Furthermore, many national politicians tend to feel that management activities within a regional organization may generate policies which go against their own national policies; they often fear the creation of a supranational body they may not be able to control individually. This fact, coupled with the fact that regional bodies may also compete with national projects for scarce financial resources, tends to create friction among national members of a regional organization, and may bring about its demise. These are, in part, some of the reasons why the East African community collapsed, after almost 80 years of existence. These weaknesses notwithstanding, regional organizations have a significant role to play in coping with problems related to environmental change, in particular those related to climate change and climate variability. Policymakers must make a deliberate effort to encourage existing regional organizations to include environmental change on their agenda.

Key Points

- No nation, however rich or powerful, can singlehandedly resolve truly global environmental problems such as stratospheric ozone depletion or global warming.

- It would be economically more efficient, in terms of manpower and financial resources, for countries in a region to pool existing resources and to encourage each country to focus on a specific aspect of a regional environmental problem, sharing the results with other states in the region.

- By pooling human resources drawn from a wide variety of countries, regional organizations tend to be better staffed, offering an optimal way of utilizing meager human resources in the region.

The South Asian Association for Regional Cooperation (SAARC) and Climate Change

Saleemul Huq

At least three types of regional organizations can be identified in South and Southeast Asia: 1) Regional organizations defined in terms of the management of specific resources (e.g., river commissions, regional seas organizations); 2) Economic-political groupings (e.g., ASEAN, SAARC and, in other regions, EC, OAS); 3) Regional funding organizations, usually regional development banks.

The environment is one of the areas in which there is widespread consensus on the major issues and approaches to be taken. This is particularly true in the area of international environmental issues such as global climate change. Probably, SAARC's most important link to the climate issue is the likely increase in hydrologic disasters resulting from climate change. However, in order for them to convert their concern and agreement on regional and global environmental issues into concerted action, both in their own region, as well as in international negotiations, a great deal more concentrated effort and support would be required.

Because SAARC is still at an early stage in its development, increasing cooperation both within the region and with other regional organizations with experience will be required.

Key Points

- Concern about the possibility of global warming may provide an impetus for regional organizations to consider regional hydrologic issues, especially as the reliability of information about potential climate change on regional water resources improves.

- In the South Asian region, conflicting relationships have long existed between states. The main achievement of SAARC in its first few years has been the very formation and survival of the organization in the face of such a considerable history of regional hostility between its various members. SAARC is now quite well established and all members feel the need to strengthen it further.

- Despite the belief that environmental issues might be less controversial, consensus has still been limited, and this lack of consensus extends to the climate change issue as well. In any event, lack of financial resources has precluded action by regional groupings on environmental issues.

The Roles of IGOs in International Environmental Management: Arena or Actor?

Arild Underdal

International organizations can serve at least two major functions in international environmental management: that of being an *arena* for the exchange of information, discussion and decisionmaking, and that of being an *actor* in the policymaking or the policy implementation process. To qualify as *actor*, an organization must (also) provide independent inputs into the policy process, or somehow amplify the outputs of the process. To meet these requirements, an organization must have a minimum of internal coherence (unity), autonomy, resources, and also in fact engage in some relevant external activities. *Arenas* differ in terms of, inter alia, *rules of access, decision rules* and *rules of procedure.*

It is important to clarify to what extent and how different rules and arrangements affect the capabilities of organizations-as-arenas to fulfill certain critical functions in the decisionmaking process – including those of (a) providing actors with incentives to adopt and pursue a "constructive," problem-solving approach; (b) providing procedural opportunities for transcending initial constraints (e.g., by coupling or de-coupling issues); and (c) institutional capacity for integrating or aggregating actor preferences.

Although *regional* organizations are primarily instruments for coping with regional problems, I should like to point out that they can serve as important arenas and actors for *global* problems. Specifically, a regional organization can influence global negotiations by facilitating coalition formation and serving as an instrument for defending or promoting the interests of its members, or by helping its members work out a joint program of contributions to a global program.

Key Points

- By providing an opportunity for discussion and decision, an organization may affect the political agenda of its member states. The mere fact that a conference is about to take place can serve as a stimulus for policymaking at the national level. Moreover, the fact that certain measures are being (or are going to be) discussed, may affect expectations of governments and societies.

- The distinction between organizations-as-*arenas* and organizations-as-*actors* does not imply a ranking in terms of *importance*. Institutions can shape outcomes as much by coupling actors, problems and occasions for decision, and by determining the rules of the game, as by entering the game as more or less significant actors.

- The best organizations for developing regulations are those whose activities caused the harm in the first place. A focus must be placed on those organizations that have direct links to desired actions and outcomes.

Role of Regional Organizations in the Context of Climate Change

Naresh C. Singh

Not all international organizations participated to the same degree or with the same interest in the negotiations leading to the Framework Convention on Climate Change. Prior to the recognition of climate change as an issue for concern, regional organizations were already undertaking a number of activities with respect to climate.

A limited review of regional organizations reveals some general problems: requirement for unanimous decisions, even when the populations of member states differ greatly in size; regional decisions made in spite of lack of national machinery, people, or other resources to implement them; and lack of national commitment to regional agreements.

In some cases, regional organizations may appear to be unsuccessful because their activities are measured against unrealistic expectations. It is not clear that climate change has special features that make it different from past problems with which regional organizations have dealt. The issues are generally not new, merely more complex.

While few existing regional organizations have been explicitly concerned with climate variability or change, many have considered other regional environmental issues. There are now a number of new organizations which focus on climate change (e.g., AOSIS).

Key Points

- A limited review of regional organizations reveals some general problems: requirement for unanimous decisions, even when the populations of member states differ greatly in size; regional decisions made in spite of lack of national machinery, people, or other resources to implement them; and lack of national commitment to regional agreements.

- The actions of regional organizations to date may in general be described as "bark and no bite." These include: mobilization of information, research, and resources; training and technical assistance; consensus building (and sometimes coordination).

- Regional organizations with a more interdisciplinary approach to decisionmaking may be required, as this is not a characteristic of existing regional organizations.

PART II

Summary of Discussion Sessions

Summary of Discussion Sessions

Michael H. Glantz and Martin F. Price

The following section contains highlights of discussions that took place during the four-day NATO Advanced Research Workshop. The process for constructing this section was as follows:

Following each thematic session during the meeting (please see Agenda in Appendix), there was discussion related to or stimulated by the theme of the session. The following section is based on the rapporteur's notes of these discussions, which identified major themes that had appeared throughout the four days of discussion. These included comments on issues related to science, policy, society, research, regional organizations, and nongovernmental organizations. The rapporteur's notes were combined with the notes of the coordinator to produce a narrative on issues related to each of the themes raised during the discussion sessions.

Science Issues

Governments now realize that global and regional climates change on all time scales. They also recognize that human activities such as the burning of fossil fuels or, more generally, the emission of greenhouse gases can alter the chemistry of the atmosphere, atmospheric processes, and climate. As a result, an increasing number of governments have chosen for a variety of reasons to support national policy actions to restrict CO_2 and other greenhouse gases emissions. As we may already have committed the atmosphere to some degree of global warming, early detection of natural and anthropogenic change is necessary, at the regional as well as global scale, and identification of possible responses to climate change are of critical importance.

The 1992 report of the Intergovernmental Panel on Climate Change (IPCC) projects global temperature increases that are somewhat lower than those cited in the first IPCC Working Group I scientific assessment issued in 1990. Recent IPCC reports suggest an increased convergence of some regional scenarios of global warming in the major general circulation models (GCMs) (especially the coupled ocean–atmosphere models); for example, the relative drying scenarios are common for the Mediterranean Basin and mid-continents, and intensification of monsoonal circulation in Asia. With regard to assessments of the regional impacts of global warming, policy responses should focus on regions for which GCM outputs appear to be in general agreement. In addition, regional climate change impact assessments should focus on those regions where different methodological approaches to climate change scenario development suggest similar types of change. Yet, regional models, like GCMs, require an improved understanding of underlying processes and interactions at different spatial and temporal scales (i.e., more basic focused research is needed). As a result, improvements in regional-scale climate modeling (including realistic feedback mechanisms), scenario development, and climate sensitivity analyses are top priorities for the IPCC.

Some scientists have suggested that, with limited financial resources, GCMs may prove to be the best current source for scenario development. They also suggest that it is necessary to continue to use different GCMs and parameterizations, because if those parameterizations currently considered to be the "best" were used exclusively too soon, some counterintuitive regional consequences ("surprises") of climate change might not be identified. The limited reliability of climate change scenarios for purposes of policymaking should be explicitly recognized. Moreover, climate change scenarios may not be reliable when they are imposed on societies that are constantly undergoing changes.

With regard to other approaches for climate change scenario development, extrapolation from past or present conditions to the future (even a few decades into the future) must be carried out very carefully and interpreted with great care. Observational data, both climate and crop yield time series, describe recent climate variability and are essential for testing models and validating their results, especially at the regional scale. Because human activities are clearly affected by extreme meteorological events, recent extreme events such as severe droughts and floods could serve as analogs for developing first approximations of societal responses to potential effects of global warming on ecosystems and societies. Such approximations could provide useful insights for the development of appropriate policy responses to climate change. In the meantime, the international meteorological and hydrologic communities have a responsibility to identify and assess the vulnerability of likely locations of serious impacts of regional water resources deriving from climate change, especially with regard to extremes.

Hydrologists and atmospheric scientists have noted that, in addition to higher atmospheric temperatures and evaporation rates, global warming is likely to result in an intensification of the global hydrologic cycle. Some regions will become wetter and others drier. For instance, projected decreases in precipitation amounts and increases in evaporation rates in such areas as the western USA and the Lake Victoria basin in

East Africa would likely result in decreases in runoff. Initial assumptions have been that climate change would lead to decreased lake levels and water quality. For example, even without a decrease in precipitation, research in the North American Great Lakes region shows that, with an increase in temperature, water quality would deteriorate. The rate of such deterioration in water supply and quality and the extent to which they become critical to ecosystems and societies will depend in part on the levels of societal demand as well as on the efficiency of use.

Increase or decrease in the flow of rivers affects river basin ecosystems, estuarine resources at the mouths of rivers, and offshore marine resources. For example, the discharge of the Amazon River affects the productivity of the Guyanese Shelf as well as nearby mangrove forests. While there has been increased attention on habitat protection and on the importance of estuaries for offshore resources, the latter has not yet led to new regulations. At such broad transboundary spatial scales, there is a need for innovative institutional research (e.g., legal frameworks and economic incentives) for water resource management.

Although it is generally assumed that we know a great deal about climate variability and its impacts on ecosystems and society, this is usually not the case. The location of these impacts and their degrees of severity vary in space and in time. For example, the same intensity of drought in different years will likely have different impacts. Thus, a critical followup of impact assessment is to determine how, in different contexts, society might allocate its resources so that the impacts of that climate variability (and extreme events) become less harmful or at least less disruptive. Understanding the spatial linkages between climate variability in one area and impacts in others will also be crucial.

Forests are important sinks for carbon dioxide, so widespread reforestation has often been recommended on a global basis. While reforestation might provide only a relatively small and partial solution to global warming (compared to control of fossil fuel use), it would have additional benefits, particularly with regard to the restoration of considerable areas of degraded land with potentially fertile soils. Vegetation, especially forests, also plays a critical role in the hydrologic cycle, although qualitative assessments of land use changes on the regional and local hydrologic cycle are still uncertain. Understanding how choices, such as whether to use land for either forests or agriculture, can have implications for national and regional effects on global warming and response strategies is a major scientific issue.

As for the impacts of climate change on fisheries, the prospects of global and regional climate change have only recently become a salient issue among fisheries scientists. Attention generally focuses on possible changes in regional abundance and distribution of living marine resources. One approach for identifying possible ideas about how societies might be impacted by changes in access to fish populations is to review the history (successes and failures) of recent fisheries.

Whereas many fish populations are sensitive to variations in climate, pelagic (surface dwelling) fish appear to be an effective biological indicator of climate-related environmental change. As one example out of several impacts of climate variability already documented, it is recognized that ENSO events (El Niño–Southern Oscillation, popularly referred to as El Niño) dislocate tuna stocks in the South Pacific. Research on climate change impacts on fisheries is starting in the Northeast Pacific and the Northeast Atlantic, but the impact of these research findings on fishery institutions and resource allocation has been minimal to date.

Policy Issues

The likely success or failure of environmental policies can best be determined on a case-by-case basis, depending on several factors. Success or failure could in most instances take many years to determine. For example, about 20 years passed from the first international concern about climate change until the development of the Framework Convention on Climate Change. Similarly, soon after the IOC was founded in the late 1960s, it established, in cooperation with the Scientific Committee on Ocean Research (SCOR) of the International Council of Scientific Unions (ICSU), a Committee on Climate Change in the Ocean. It was many years, however, before the World Meteorological Organization (WMO) and the IOC developed joint programs; and as yet there has been little interaction to date between WMO and IOC activities at the national level.

Six lessons may be drawn from these examples: 1) the gestation period of an international policy occurs primarily during the technical phase, rather than the policy phase; 2) the duration of the gestation period depends in large measure on its perceived importance by various affected nations; 3) regional programs on climate change must take into consideration as well as involve those groups that perceive themselves to be affected by such change; 4) concerned scientists must convince their own research establishments and governments to act on critical issues; although technical inputs are necessary to the policy process, they are not sufficient for policymaking; 5) in the context of global warming, it would be a fundamental mistake to link scientific research and regulation functions within the same regional organization; and 6) the time period over which observable changes in global or regional climate are likely is on the order of a few decades or so, a time frame compatible with, and directly relevant for, the development and construction of major industrial projects such as dams. This is a longer time frame than the tenure in office of most public officials. Therefore, governments must develop ways of dealing with this mismatch of time frames as well as with intergenerational issues.

Some key policy factors that must be considered in assessing the potential role of regional organizations in the context of global warming include, but are not limited to, the following: 1) the political context of change: today's political context and the capacity to take the required actions, have changed markedly during the recent past; 2) the type of environmental problem: if rule-making had obvious benefits, the speed at which resolution might be achieved can be quite high (perhaps 7 years), unless there is a liability convention (9 years). If the climate-related regional environmental problem is complex, and the potential solution will likely result in costs, the process leading to action will likely be lengthy. In many cases, the rate of environmental change and the response time by decisionmakers are out of phase. Time is needed for learning and for involving new organizations and putting them together in new groupings.

Knowing when to act – when the amount of scientific information is enough for political action – is an ever-present problem for decisionmakers. Action taken by policymakers to adapt to new political as well as environmental changes can be delayed for a long period of time, if not indefinitely, in the name of reducing scientific uncertainties, even though enough information may be available to warrant at least precautionary action. However, policy decisions have to be made in spite of remaining scientific uncertainties, because the policymakers and the public cannot wait indefinitely for the possible errors surrounding climate projections to be reduced by future research findings. Causes and effects of global warming will in many instances be regional, hence also will be the necessary responses to its impacts. Climate change can be linked to many regional and national economic development issues. In fact, it can be viewed as *the ultimate* environmental issue, hindering and perhaps foreclosing the attainment of sustainable development in many or perhaps all regions. In a period when net transfers such as foreign aid to developing countries are decreasing, emerging climate-change-related environmental issues in the Third World provide an increasingly important and valid reason for renewing sorely needed financial and technical assistance from industrialized countries. No less important is action by the industrialized countries to reduce the causes of climate change.

The reliability of predicting possible tactical and strategic responses by governments to regional problems decades in the future is greatly limited by unforeseeable institutional as well as demographic changes that take place within societies; by the degree of scientific consensus at a given point in time on global warming and its regional consequences; and by varying national or sectoral perceptions of who might win and who might lose as a result of the regional impacts of climate change. One prevalent view suggests that governments tend to respond only to crisis situations. But crises and societal responses to crises are very difficult to predict. Thus, political and social responses to climate-related environmental changes will tend to be discontinuous and attention likely to be limited, except during episodic events.

The dislocation of certain kinds of resources as a result of global climate change (e.g., water resources, fish populations, forests and agricultural production) are most likely to occur at the regional level. Thus, the threat, if not the reality, of climate change could provide a catalyst for regional cooperation when countries within a region realize that transboundary changes in the availability (e.g., location or abundance) of their resources require regional cooperation for the rational management of those resources. Interstate cooperation would be necessary for developing regional strategies to cope with climate-change-related resource impacts, as well as ways to deal with changes in bargaining power vis-à-vis various regions.

Control measures to limit the emission of greenhouse gases have to start at national and regional levels. The initial goal of the Framework Convention on Climate Change to stabilize the level of atmospheric emissions of greenhouse gases may appear to those seeking more aggressive actions to be among the least ambitious actions. Nevertheless, it will be the first step necessary to get governments on the way to more effective measures. Once committed to stabilization policies, governments would come under pressure to take more aggressive actions, if warranted by incoming scientific information. The ultimate goal of the Convention, to stabilize atmospheric concentrations, will require extensive additional commitments to reduce emissions by 1–2% per year over a number of decades. To identify appropriate control measures, there has been a strong push for the essential surveillance and monitoring of the atmosphere and oceans necessary to assess trends. The assessment of national and regional trends in greenhouse gas emissions requires an objective international scientific capability, such as the IPCC Working Group I's scientific assessment, divorced from direct political influence.

While it is not yet clear what protocols to the Convention will be necessary, the convention itself contains only limited commitments. It is risky to draw too many parallels with the convention and protocols on ozone depletion. In the latter case, the potential impacts were regarded seriously by governments because of public concern about direct health effects of ozone depletion, and because CFC major manufacturers were worried about liability as well as market share, among other key factors.

The position of the US government in the Climate Convention negotiations changed during the process, and is not as deeply rooted in the federal bureaucracy as compared to many other environmental issues. Moreover, powerful industrial interests (e.g., public utilities, coal, oil, automobiles) oppose restrictions that the Convention would foster. Economic, fiscal, and regulatory incentives may be needed if these industries are to change, as would be necessary for the implementation of the convention.

For their part, Russia and China rely heavily on coal for energy. Their dependence on fossil fuels for development presents major difficulties for compliance with the Climate Convention. The transfer of energy-efficient technologies would go a long way toward reducing their dependence on fossil fuels for implementing their economic development programs. However, foreign assistance and technology transfer related to the reduction of greenhouse gas emissions face a number of barriers and raise issues of intellectual property rights and trade. Because in many countries "command-and-control" remains the dominant politico-economic approach, diffusion of technology and resources are also hampered. This could either help or hinder implementation of the Convention.

Without large-scale financial resources or innovations in international law, it will be difficult to develop regional cooperation. Some aspects of cooperation (e.g., standardization of measurement), however, do not necessarily require significant financial resources. In general, one could argue that political will is more critical to success than are financial resources; all parties must see a benefit, financial or otherwise. In developing regions, where there is policy agreement among states, regional projects may attract funding more easily than national ones. The funding requirements for a regional organization depend on its final objectives. For example, operational phases (research, monitoring, data analysis, etc.) inevitably require long-term, uninterrupted financial support.

The original focus of negotiations on climate was on a Convention for the Protection of the Global Atmosphere, not on climate change per se. The emphasis then shifted to climate change. The existing intergovernmental structures (IPCC, INC, Framework Convention, etc.) are proceeding on the basis of incomplete scientific research and observations. "Incomplete" is a better characterization than "limited," considering that the monitoring and research effort and budgets are already quite large, and that the models play a more important role than the observations in highlighting the risk of global warming. However, little in the way of additional funds have been found for improving observation systems or increasing understanding of the interactions between hydrologic and biological systems within the climate system. The scientific understanding of the climate system (in the broadest definition) is crucial now, because there is a need for general understanding and common diagnosis of a problem before effective norms and rules can be formulated to address that problem.

The effect of land-based pollution on streams, estuaries and coastal zones is increasingly recognized as an issue for fisheries management, particularly for near-shore fisheries, as 70% of all marine pollution is near-shore. Research is needed on the effects of pollution on these sensitive ecological systems, particularly for juvenile stages.

Recognition of global warming (natural or anthropogenic) and its likely impacts implies that the degradation or alteration of environmental resources such as the atmosphere (or climate, more specifically) can no longer be treated as being without value (e.g., as a free good). The development of methods to evaluate the costs of using the environment (in this instance, the atmosphere) is necessary. However, an objective determination of accounting for environmental degradation at the national and international levels is a very complex task.

Specific regional policy-related examples

- As most GCMs agree in projecting that aridity is likely to increase in the Middle East region, it is urgent to resolve water allocation issues and improve climate change predictions for the region. However, climate change has not yet become an issue in that region for policymakers and water resource planners.

- Although total water supplies may be adequate, the situation in South America is similar to that in the Middle East with respect to water distribution and allocation problems, especially for hydropower production. In the Latin American situation, regional and intergovernmental organizations are involved in planning and assessing the costs of the use of shared resources.

- Frequently, regional and national institutions that develop strategic plans for natural disaster preparedness (i.e., natural events such as floods or droughts that might be affected by climate change), often do not take into consideration the possible implications of climate change (e.g., the Bangladesh government's Flood Action Plan).

- While there have been innovations in the management of multi-species fisheries, not all have been successful, e.g., cod stocks in the North Atlantic and West Coast salmonids. In addition, while major theoretical advances have occurred in the European Community (EC), the results are less impressive in terms of enforcement by member states, in spite of a centralized authority. Constraints have included the active collusion between states and the lack of reliable statistics.

- European inland fisheries need to be more seriously considered in water resources management agreements. This need was highlighted by Romanian releases of pollutants, without warning, into rivers that flow into Hungary. As another example, in Hungary's Lake Balaton, competition between sport and commercial fishermen is growing. In other countries, fish stocks are being affected by pollution, a changed water regime, and new dam construction. The inability of fish to traverse fish ladders is leading to a larger stock, but of species less valuable to society. This situation is being complicated by introduced species. Climate change is likely to greatly influence species composition. Reservoirs may become refuges for some species, if rivers become ephemeral and pollutant concentrations rise.

Societal Aspects

Studies by insurance companies as well as impacts researchers show that in recent decades there have been increases in economic losses resulting from "natural" disasters (especially severe storms such as hurricanes and tornadoes, as well as droughts, floods, and frost). Because the exposure of human communities to natural phenomena is increasing as a result of population growth and movements into areas of risk, it is important, if not crucial, to identify the relative contributions of natural, social, economic, and political factors to these losses.

Environmental impact assessment provides a useful model for a professional approach to identifying and analyzing the societal aspects of climate change. It is a very structured process, based on the normative principles and values of the USA's National Environmental Policy Act of 1969. Each environmental impact statement is subject to public scrutiny and sometimes judicial review. However, there is a risk of infatuation with the technique, and so it must be used (and its results interpreted) with great care.

When developing societal response options in relation to climate change, it would be misleading to focus only on the obvious primary impacts. In fact, the old unintegrated sectoral approach (forestry, agriculture, fisheries, sea-level rise, energy, etc.) could result in the development of inappropriate societal responses. Consequently, realistic response options to aggregate impacts will be far more complicated than any approaches attempted to date.

For many societies, adapting to climate-related environmental change may become more difficult because of the rapid rate and the intensification of processes of environmental change. Societal institutions may not have enough time to adapt to one apparent change before other, less obvious, ones appear. This will likely be an acute problem for many "traditional" societies. Such societies had developed effective strategies for coping with climate-related environmental variability. However, the stability and effectiveness of their traditional coping mechanisms has decreased as the linkages of these societies to the modern world increase. Under some conditions of environmental change, institutions as well as property regimes can become unstable. While it may be possible to adapt incrementally to gradual changes, it would likely be more difficult to adapt to rapid changes.

Institutions and property rights in the fisheries sector

When countries do not have domestic fleets with adequate capacity, they sell licenses for fishing rights to other countries. This is possible because of a clear property regime. However, developing countries are in a weak position in such cases, especially when part of the licensing fee is not paid by the fishing industry. For instance, the EC pays part of the license fee and compensations charged by foreign countries to authorize EC vessels to fish in their EEZs (Exclusive Economic Zones). Overfishing results because

EC fishing companies do not support the fishing cost corresponding to the value of raw materials (or harvests).

The allocation of scarce resources requires a clear property regime (including sharing of resources); clarification of the uses, and rights to use, of property; and efficient allocation mechanisms.

Examples of potential impacts of climate change on fisheries already exist. These may provide analogies for assessing how fisheries property regimes might cope with future climate-related environmental change.

The new Law of the Sea has had perhaps its greatest impacts in the EEZ as a result of the extension of coastal state sovereign rights over all living and non-living resources within the zone and the extension of the exclusive jurisdiction of the coastal state over all economic activities occurring within the EEZ. However, the immediate effect on fisheries within the EEZ has been the displacement of a significant amount of fishing effort of Distant Water Fishing Nations (DWFNs) and a significant growth of coastal state harvesting capacity. The result is that DWFNs have retreated increasingly to high seas areas, while coastal state fleets have been increasing largely in an uncontrolled fashion. The inevitable overcapitalization leads to overfishing of stocks within EEZs by coastal state fleets, which then increases the pressure on high seas resources as *both* coastal states and DWFNs seek alternative opportunities to fish.

With respect to straddling stocks, there are problems, especially in two areas: 1) closing arrangements for new entrants, when catches have been completely allocated; 2) surveillance and enforcement, particularly for those who are not parties to an agreement. With respect to shared stocks, a major problem occurs when different parties fish a stock at different times in its life cycle.

Distortions in institutional structures lead to distorted investment patterns. However, changes in institutional structures may not resolve investment patterns, while there are still profits to be made. Inertia in institutional change is a long-recognized handicap to organizational reform.

Almost as much of the world catch is discarded at sea as is landed. If fish is to be used to decrease protein deficiencies, appropriate products will have to be developed to match culturally-induced perceptions of "food." Without a market, the waste will continue, and inequities in food distribution at the global level will continue.

Other policy issues

It is important to note the relationship of the distribution of human populations and their consumption of fuels and agricultural products. "Overconsumption" in industrialized countries is as crucial an issue as population growth in developing countries. The growth of human populations and the demands that accompany increasing affluence are the primary underlying sources of increased anthropogenic GHG emissions. The population issue, however, has largely been excluded from policy debate and formulation, even at UNCED in Rio de Janeiro in June 1992, the largest environment-related political conference ever.

Some consequences of carbon dioxide enrichment and global warming (e.g., increased crop yield depends on both CO_2 fertilization and warming effects) may be beneficial to certain countries or sectors, while costly to others. Some economic sectors will "lose" from a restructuring of the energy sector in order to modify its greenhouse gas-producing behavior. Government assistance will likely be needed, at least to cover short-term costs of acquiring and implementing new energy-efficient or emission-reducing technologies. Trade relations can be used by some countries seeking to reduce greenhouse gas emissions to persuade others to do likewise, if compatible with the structure of GATT. Better approaches in trade and technology transfer emphasize cooperation, not confrontation. Regional organizations have a role to play in such activities; for instance they could collect subsidies and distribute them to energy-efficient companies.

Compensation for losses, however, may not be an appropriate strategy; for example, with subsidized insurance in the United States, people continue to build in flood-prone areas. Thus, unless legal precepts and institutional structures are changed, existing problems will be exacerbated as regional win/loss distributions change. It is important to note that many water resources management problems can be expected to occur, even in the absence of climate change.

While new research will provide better information on institutions and mechanisms that can facilitate the efficient use of the global ecosphere, policy problems will likely remain, especially as the distribution of gains and losses changes among states and regions. Thus, institutional innovations are necessary for adapting property regimes, allocation mechanisms, and use rights.

Regional Organizations

The impacts of climate variability on society and ecosystems for the most part do not respect international borders. Thus, transboundary collaboration on coping with such impacts is usually more effective, as, for example, in the cases of the southern African and Sahelian droughts, where regional groups (SADCC and CILSS, respectively) have developed regional tactical and strategic responses, both short-term (e.g., SADCC's pledging conference) and long-term (e.g., CILSS's ongoing environmental programs).

The creation of new regional organizations or the reorganization of existing ones often derive directly or indirectly from public demands on governments for improved management of transboundary resources such as watersheds, airsheds, and marine fisheries. The creation of effective regional organizations must emerge from the desire of states within a region because if they are imposed on a region by the UN or other international bodies, they are not likely to survive. Thus, the question is generally not whether, but how, to design effective regional institutions. One reason that many regional organizations are hesitant to expand their domain is that new skills and resources would be required to perform the new tasks. This often leads to resistance to changes in the scope, or in expanding the activities of existing organizations, despite the changing real needs of its members. For example, the Danube Commission is restricted to navigation and has neglected the growing concern over other factors. Thus, while it may be tempting to develop a new integrated system based on existing regional institutions, this may be risky and generate a large number of significant new problems. However, while regional organizations tend to resist innovation and change, technological as well as political changes push them toward institutional evolution. In the face of emerging global climate-related environmental issues, regional organizations that may have been ineffectual in dealing with regional or global environmental issues in the past may not necessarily remain so in the near future.

Although regional organizations differ in structure and function, they often face similar problems. This suggests possibilities for the exchange of information among them for mutual benefit. The need for such exchanges is growing, especially for global issues in general and the global warming issue in particular. Recognition of the prospect of climate change also highlights the need for environmental education to change worldviews, and for climate-related information exchange through regional organizations. This could be valuable for building on successes through the wide distribution of regional case studies of successes and failures in coping with climate-related societal stress.

The activities of many regional organizations are often driven by the most powerful countries within them. With the end of the Cold War, fundamental divisions between types of organizations are changing, such as the shift from bilateral linkages between the former USSR and other circum-Arctic countries to the regional Arctic Commission, or the emergence of a serious regional approach to addressing pollution problems in the Baltic Sea negotiations. With regard to the climate change issue, regional organizations could focus on technology transfer, research, monitoring, reporting, information exchange, and the harmonization of legal instruments.

Existing institutions all have an "essence," or sense of mission, that must be built upon. For NATO, this relates to security, defense, and diplomatic issues. Other regional organizations offer different types of solutions. For example, regional development banks offer market solutions, river basin organizations use legal and regulatory concepts, and political organizations offer principles (e.g., equitable utilization and equal access). While there are compelling reasons to use existing institutions, they must become more interdisciplinary. Unless the actions of a regional organization are geared to address all the roots of a problem, only a low level of cooperation may develop, persist for a while, and then decay. The importance of the existing "bark and no bite" role of regional organizations should not be downplayed, because rhetoric ("bark") may lead to action ("bite").

With respect to the climate change issue, three features of regional organizations make them important: 1) their political influence, deriving from the consensus of several governments, acting in common self-interest, and often in non-traditional groupings based on shared impacts of climate change; 2) their ability to mobilize financial resources (although a regional organization can permit funds to be allocated over a long period at a regional level, it is essential that member states agree on the use of those funds); and 3) their relatively narrow focus compared to global organizations. Among the roles of regional organizations are the minimization of conflicts between national governments (e.g., through consensual reports by scientists and technical advisors from all member countries).

The most effective type of regional organization for the issue of global climate change may be one that has been organized as part of a global organization or network. One appropriate example would be the International Geopshere-Biosphere Program

(IGBP) START program, an initiative to assist in the development of indigenous capacities through a network of regional institutes and centers. A second example is the WMO's regional structure, in which regional associations cooperate in such activities as cyclone warning and preparedness systems, and standardized research, data collection, and exchange.

Four important general lessons can be drawn from past experience with fisheries institutions with respect to climate change:

(1) The nature of the decision system is critical. At the global level, there is no alternative to the "law of the least ambitious program." If each country has one vote, regulations are likely to be watered down in order for them to be acceptable to all. Yet, while a general law can be suggested that "no international organization can work better than the average of its members," one could argue that even a minimal level of beneficial actions are better than doing nothing. For the climate change issue, it may be better to proceed sequentially, starting with industrialized countries, which should decrease greenhouse gas emissions first and fastest, leaving time for additional research and for economic growth in developing countries. It was suggested that the best organizations for regulation are those whose members' activities have caused the harm in the first place (e.g., the OECD, the EC). A focus must be placed on those organizations (and countries) that could have a direct impact on desired solutions.

(2) The divisibility of payoffs is critical. For example, how should costs be apportioned? How can regulations be used to ensure equity?

(3) Regulation requires monitoring and enforcement, so that the costs of cheating are much greater than the potential gains from cheating.

(4) The role of science in the debate is crucial. Some observers have argued that the only reliable scientific voice could be a global one (e.g., the IPCC). Others, however, may argue that the IPCC is suspect precisely because it is intergovernmental, and that it may have weakened its scientific credibility in seeking to arrive at a consensus too soon. Scientific data gathering and analysis must not be linked so closely to regulation that governments meddle with it – governments will tend to suppress scientific information that does not support their national positions.

While few existing regional organizations have been explicitly concerned with climate variability or change, many have considered other regional environmental issues. However, a number of new organizations do focus on the climate change issue. Some are functionally based (e.g., AOSIS, IGBP-START program), while others are global programs (e.g., IGBP, IPCC, International Social Science Council's Human Dimensions of Global Environmental Change Program, World Climate Program).

Climate change is only one factor that could affect a country's or a region's economic development prospects. While the climate system may be less sensitive to rapid increases in greenhouse gases than many may think, we have to recognize that the risks cannot be ruled out. Such risks will undoubtedly increase because of the combined effects of expanding human and livestock populations and climate change at the regional level. Regional organizations have a major role to play in the climate area, even in the absence of global climate change.

Thus, there are a number of conditions for the effective action of a regional organization: 1) specificity (i.e., focused response); 2) allowing those directly affected to decide on appropriate responses; 3) in industrialized countries, there must be significant gains to be made or significant adverse effects to be avoided, and major national political constituencies that push for action. One example is the Economic Commission for Europe's (UN) Convention on Long-Range Transport of Airborne Pollutants (LRTAP) (1979). Following this treaty, a Protocol was negotiated, the Helsinki Protocol (1985), which committed countries to a 30% decrease in SO_2 emissions and a second protocol on NO_x emissions. These efforts were supported by a cooperative monitoring program. The SO_2 program is currently being renegotiated to try to achieve greater reductions. Some "services" are better provided at the regional rather than national level so as to avoid duplication of effort and multiplication of cost, for example, of research, use of limited human resources, and attracting funding from outside the region. This is important in relation to the growing number of new institutes for global environmental issues. Who will staff them, and how should responsibilities be distributed between national and regional organizations? Such institutes are easier to create (usually for political reasons) than to maintain and keep efficient. Even within a country, different agencies tend to pursue their own, often conflicting, foreign policies. The resulting short-lived institutes and networks often complicate issues, making long-term coordination difficult.

Although there can be advantages for involving non-UN regional organizations in UN activities (such as the IPCC or the INC), there are a number of problems:

usually such organizations are given only observer status; they lack information about operations within the UN system; and there are limited possibilities for funding their participation. However, the UN itself can provide catalytic funding for regional organizations. If UN conferences were to promote the cause of regional organizations, resources would have to be found for long-term support of those organizations.

One recent UN activity, UNCED (and the process leading to it), has increased interactions between regional organizations. Regional organizations could play an important role by assisting developing countries to prepare inventories of sources and sinks of GHGs and to develop policies which explicitly consider the regional causes and potential regional effects of and responses to global warming.

One of the reasons for a possible increase in the significance of regional organizations, especially those composed of scientists rather than government representatives is that they could make strong statements that specific national governments are not able to make.

Time is needed for learning and for involving existing, or creating new, regional organizations in the global warming issue. This always presents a dilemma, as action can be postponed indefinitely by claiming an institutional need to adapt to new scientific information on climate-related environmental changes.

As a final comment, many existing regional organizations are well-placed to contribute to the development of policies relating to global warming, and some regional organizations were active in the development of the Framework Convention on Climate Change.

Some regional examples cited during the workshop

- The Plata Basin Treaty was signed in 1967 by Argentina, Bolivia, Brazil, Paraguay and Uruguay. This basin and its area of influence generates 80% of the GDP of the five countries. The coordinating body of the Treaty is the Intergovernmental Coordinating Committee (CIC), based in Buenos Aires. Working groups composed of experts from the basin countries consider specific topics such as water quality and other water-related issues. Recently, the CIC has been restructured to become an Executive Secretariat for the Plata Basin Treaty. Changes in the use of natural resources of the Basin, especially land and water, have exacerbated several problems: erosion of productive land, soil and water contamination, floods and other natural hazards. Some extreme floods are caused by natural phenomena such as ENSO conditions. The rainfall regime of the river basin, its agricultural development, and the presence of numerous dams all have an impact on watershed activity. The series of large floods that have devastated the basins of the Paraguay and Paraná rivers in the last several years, warrant a detailed study to analyze their origin so that preventive measures can be taken.

- The International Joint Commission (IJC) for waters along the border between the United States and Canada was established by the 1909 Boundary Waters Treaty. This treaty primarily focuses on water levels, apportionment, and navigation, but also includes a pollution provision. Although the latter was not regarded as important for decades, it eventually provided the basis for a 1972 water quality agreement for the entire basin of the Great Lakes. From the 1970s, the functions of the IJC have increased, with a growing focus on ecological issues. Its six members, political appointees, provide, with support from scientific advisory boards, both scientific and policy advice to Canadian and US government agencies. These bureaucracies are at times turf-conscious and conservative, and are worried about proposals which may compromise their authority. In the early 1990s, both Canadian and US government agencies began to decrease the authority of the IJC, with an increasing amount of IJC work being carried out by national agencies. There have been various suggestions either to create a new body or abolish the IJC.

  One specific example of the constructive role of the IJC is the Garrison Diversion Project. The Garrison Diversion was planned by two federal agencies, respectively responsible for the upper and lower basin in the Dakotas. The diversion project would provide water for irrigation of land previously used for dryland farming. The irrigation return flows would continue to Canada in the Red and Souris Rivers. The project soon became politicized in part by Canadian NGOs, based on their concern about potential increases in water pollution by agricultural activities, and concern about the impact on the Lake Winnipeg fishery that would result from the migration of fish from the Mississippi watershed. The concerned Canadian NGOs developed linkages with major US NGOs at the national (but not the state) level. At the request of the governments, the IJC conducted an investigation which resulted in subsequent United States modification of the plan so that all return-flow water stayed in the Missouri basin. The general lesson to be drawn from this example is that all parties must be consulted from the outset of the planning process, and that existing treaty arrangements can be used to avoid environmental conflict.

Other river basin examples briefly mentioned during discussion sessions are the following:

- The Niger Basin Commission was created by the eight riparian governments. It has subsequently fallen apart, leading to the severe decline of a hydrological monitoring and forecasting project developed by WMO and UNDP. Something similar occurred in the Lake Victoria basin.
- In the Nile basin, there is some basin-wide cooperation on monitoring projects, although no regional organization for water resources has been formed. Regional demands for hydropower and irrigation are increasing, but there is currently little development in Ethiopia, southern Sudan, and Uganda, and projects there could have adverse effects on Egypt.
- The impact of global warming on inland lakes may become critical in terms of both water quantity and quality.
- A new project on the environment of the Danube River deserves mention, as a new activity supporting two new treaties on the Environment and on Water.
- In the European Communities (EC) recent treaties have given increasing attention to environmental issuès, such as those related to water resources. It appears likely that the river basin concept will be adopted throughout the EC, with common rules throughout the region. On the Iberian Peninsula, existing binational agreements are very limited, considering only hydro-electric power on rivers but not watersheds. New EC policies, however, may lead to cooperation on water quality.
- In the Ganges basin, there are only bilateral agreements (India–Bangladesh, India–Nepal), as India has opposed the creation of a basin-wide commission with real powers.
- The Lake Chad Basin Commission was created by the Lake's four riparian nations (Nigeria, Niger, Chad, Cameroon) to deal with mutual claims on water resources for navigation and trade and, later, environmental concerns. The creation of the commission was hopeful and significant but, because of the lack of will of member states, has not even come close to realizing its potential.
- The Treaty for Amazonian Cooperation (TAC) was signed in 1978 by the Amazonian countries: Bolivia, Brazil, Colombia, Ecuador, Guyana, Peru, Suriname, and Venezuela. It provides for collaboration among the member countries to promote

scientific and technological research and exchange of information; protection of navigation and trade; preservation of the cultural heritage; health care; and the creation and operation of research centers. The TAC seeks to promote the integrated development of the basin to improve the standard of living of the local population and permit the conservation and rational use of their natural resources. Meetings of ministers of Foreign Affairs of the contracting parties are held periodically to set the guidelines for common policy and evaluate the overall process. The Council on Amazonian Cooperation considers initiatives presented by the countries and makes the appropriate decisions on studies and bilateral or multilateral projects to be implemented in the framework of the TAC. The Treaty Secretariat, whose functions are currently carried out by whichever country hosted the most recent Meeting of Ministers, is responsible for carrying out the activities provided for in the Treaty and those mandated by the Meeting of Ministers of Foreign Affairs. The TAC member countries are considering transforming the Secretariat pro tempore into a permanent Executive Secretariat. Special Commissions are created to study specific problems or issues relating to the purposes of the Treaty. To date, six special commissions have been established: Environment; Science and Technology; Health; Indian Affairs; Transport, Infrastructure and Communications; and Tourism. They receive the support of international and regional agencies in formulating and implementing their programs and projects. The OAS is collaborating with the TAC countries in the implementation of the following binational studies on border integration and development: Colombia–Ecuador, Colombia–Brazil, Colombia–Peru, Brazil–Bolivia and Brazil–Peru.

- OAS multilateral technical cooperation emphasizes integrated studies in plurinational projects either in border areas or in multinational river basins. Such projects facilitate communication within the region on conservation issues and can provide a basis for common approaches to management and control of natural resources along common borders. More important, multinational participation strengthens cooperation among national governments and can lead to multinational agreements in resources management. A great need of OAS member states is a regional system to monitor and report on compliance with treaties to which they are signatories. A regional agency can play an important role in this respect. A regional approach to environmental problems also attracts more technological and financial assistance, when and where necessary, than individual countries can secure individually.

- At the 1992 International Conference on Water (Dublin), no consensus emerged between those downstream countries which desired regional organizations for water resources management and upstream countries which did not want formal references to shared water resources. There are also, however, examples of upstream countries whose development options have been restricted because of powerful downstream nations (e.g., Ethiopia/Egypt, Nepal/India).

Non-Governmental Organizations (NGOs)

As new environmental issues emerge, existing regional organizations may prove to be either unable or unwilling to deal with them. Thus, NGOs – particularly the regional and global ones – could have an important role to play, and could assist in the development of new regional organizations as they now do with international environmental policies.

NGOs and regional organizations have several similar characteristics. However, care must be taken not to overstress similarities between these two types of organizations because major differences also exist.

Many NGOs are regionally based and have been focused on regional (as well as global) environmental issues, many of which are climate-related. Only recently have the concerns of many regional organizations changed to include environmental issues, with a relative decrease in emphasis on their original functions (e.g., economic, political, navigation).

NGOs are participating increasingly in the process of transnational policy development. Through their participation, NGOs can help to legitimize the resulting policies. However, there is a great diversity of tactics and objectives of NGOs and they can be played against (or can compete with) each other, thereby neutralizing NGO influence.

Environmental and industrial NGOs have devoted an increasing portion of their resources (time and money) to influence government policies. NGOs operate at various levels of social organization from the local to the national to the regional to the international level. There appears to have been an increase in the risk of deliberate distortion of facts or of selective inattention by advocacy groups (e.g., not paying attention to factors that might contradict the group's position), especially by those seeking to gain widespread media coverage of their views. As a result, policies pursued by some NGOs are based primarily on political objectives and not necessarily on scientific information.

NGOs can be seen as counterweights to government bureaucracies. They are not particularly well liked by most government agencies. NGOs, however, do tend to reflect social currents. Many of the issues which NGOs put on environmental agendas in many cases might not be put there by government agencies because of domestic or international political sensitivities.

In recent years, there has been a major growth in the number and activities of NGOs in developing countries. These are often led by physical scientists and researchers. Sometimes they are tolerated even by repressive regimes. There are also NGOs led by former government officials but, in many instances, these lack the widespread support of the public.

NGOs have a particular role to play in the growing movement to address intergenerational and other issues that have a time scale longer than the tenure as well as interests of national governments.

Given that the frequency of such conflicts (or opportunities for cooperation) could increase in the face of global climate change and its regional consequences (e.g., the shifting of highly valued resources), increased attention should be given to the role of regional as well as nongovernmental organizations (both existing as well as those that might yet be established) in resource-related regional conflict resolution.

PART III

Discussion Papers

BACKGROUND ON THE CLIMATE CHANGE ISSUE

The Regionalization of Climate-Related Environmental Problems

Michael H. Glantz

During the 1950s and 1960s, regional organizations received considerable attention as potential contributors to cooperation and peace around the globe. At that time, Joseph Nye wrote about the regionalization of world politics, capturing the "spirit of the times" in his book, *Peace in Parts.*[1] Citing others, he noted that "the political life of the world is becoming less at the level of national states and more at the level of continents." It appears, however, that interest in regional organizations declined during the 1970s and 1980s compared to that given to nations and the United Nations.

With the end of the Cold War and the decades-long rivalry and competition between the US and the USSR, there was considerable hope that the United Nations system would be able to operate as it was originally intended – as a community of nations fostering cooperation and discouraging aggression. That hope was recently renewed as a result of the Persian Gulf war, when UN Security Council members apparently worked together (however reluctantly) to reverse the consequences of an Iraqi takeover of Kuwait. For a moment, such unity gave the impression that international cooperation (at least among the major powers) would become the rule, and not the exception. However, it appears that euphoria may have been somewhat premature, given the numerous conflicts presently under way in all parts of the world, even in Europe.

Perhaps the 1990s will provide a "window of opportunity" for an objective assessment of regional organizations and their potential contribution toward resource-related cooperation and conflict resolution. Given the resurgence of worldwide concern about "global" environmental issues and their regional causes or consequences, regional organizations could provide an effective arena for discussing, resolving or averting regional conflicts related to environmental change. It is useful to note that natural or anthropogenic environmental problems could be viewed as global for a variety of reasons. They could be global in cause (e.g., the onset of an Ice Age); they could be

[1] Joseph S. Nye, 1971: *Peace in Parts: Integration & Conflict in Regional Organization*, Boston: Little, Brown & Co. (Reprinted in 1987 by University Press of America.)

global in effect, although local in origin (e.g., ozone depletion, global warming, tropical deforestation); or they could be local in cause and effect but of worldwide interest (e.g., desertification). The nature of a problem defined as "global" will often determine how it is approached by the community of states.

There has been considerable discussion of and, lately, debate about the prospects for the occurrence of a global climate change on the order of a few degrees Celsius during the next several decades. Although the consequences of global climate change will be (or, as some scientists now argue, are being) felt at the local and regional levels, climate modelers are not yet in a position to provide societal and environmental impacts researchers with the regional detail needed for the construction of reliable scenarios. Thus, national and subnational policymakers have become increasingly concerned about how a global climate change might manifest itself at the regional and local levels. The time may be right to talk about the "regionalization of environmental problems."

To date, most attention has focused on the causes of global warming. This has, in turn, sparked concern about the policy options countries might pursue to prevent, mitigate, or adapt to the consequences of global warming. After a long and arduous negotiating process, this concern has generated the Framework Convention on Climate Change.

Attention has also turned to developing a variety of methodological approaches to gain a glimpse of the possible regional consequences of a global warming. To date, most scenarios about the potential regional effects of global warming have been based on the outputs of various general circulation models (GCMs). Yet, because the precipitation scenarios produced by different GCMs disagree substantially for individual regions, their use for developing policy prescriptions is risky. In addition, GCMs provide little in the way of reliable information about likely changes in the location, frequency, and intensity of extreme meteorological events. Questions about the potential severity of impacts pertain not only to climatic factors (including extreme events), but also to their relationship to societies possessing different capabilities for implementing proactive as well as reactive policy responses. These differences with regard to response capabilities raises the issue of climate-change-related "winners" and "losers."

In addition to a reliance on GCMs for developing plausible scenarios of likely future regional climates, other methods have been used:

- Paleoclimatic reconstructions: Past climatic periods, such as the Altithermal (4000 to 8000 years ago), may provide analogs to the expected climate of the next century. However, the forcing mechanisms that produced the atmospheric warming were quite different. Paleoclimatic reconstructions, like forecasting by analogy approaches, must be used with caution.

- "Warmest arctic summers": This approach is based on the fact that GCM scenarios project that atmospheric temperatures would likely be notably higher in summer at high latitudes than at midlatitudes. Consequently, the global weather patterns of a set of previous years in the 20th century with unusually warm arctic summers were identified and used to suggest possible climatic patterns in the midlatitudes under global warming conditions. The reliability of this approach to regional scenarios development has been scientifically questioned.

- Historical 20th century analogs: Sets of decades throughout this century have been warmer (1970 to present) or colder (1940–70) than average in certain regions. Once these decades have been identified, the responses of both ecological and societal systems during these periods can be examined to assess the regional climate sensitivity to decadal-scale climate fluctuations.

- "Forecasting by analogy": Societal responses to recent climate-related environmental changes (e.g., lake level rise in the Great Lakes, the Salt Lake, and the Caspian Sea, and decreased river flows) may be examined to identify weaknesses and strengths of present-day institutional responses. Such studies have, for example, suggested that rates of environmental change are often as important to decisionmakers as processes of change. There are benefits to be gained by learning about how well different societies cope with environmental change.

In sum, it is important to select with care those scenarios that may be used to define possible policy actions at the regional level. One must also recognize that regional counter-intuitive changes cannot be ruled out.

Most likely, the impacts of a climate change on society will be directly felt as changes in the variability of climate, and more specifically as changes (or perceptions thereof) in the frequency, intensity, duration, and location of extreme meteorological events. Even with regard to adverse impacts of changes in the "global commons" (the atmosphere, the oceans, the tropical rainforests), the regional impacts of those changes will most likely manifest themselves in different ways and, perhaps, even in different directions.

For example, the citrus-growing region of the US state of Florida witnessed at least five major freezes in the 1980s, the hottest decade on record. This example suggests that regional counter-intuitive surprises should be expected with a global warming of the atmosphere. It is possible, for example, to gain a glimpse of how well prepared societies may be for changes in the environment by looking at recent ecological and societal responses to environmental change in different parts of the globe.

How then might regional organizations face an uncertain climatic future? Even in the absence of a CO_2-induced global warming, the global climate regime fluctuates on all time scales of interest to society. How might decisionmakers pursue sustainable development policies during periods of climate-related environmental changes?

Whereas regional international organizations (functional as well as geographic) have often been relegated to marginal roles in the international political arena with regard to resources issues, this will likely change in the future. As resources and people dependent on them "migrate" on land and in the marine environment, the risks of regional conflicts, as well as the opportunities for regional cooperation, are likely to increase. This can be expected because, as climate regimes shift in response to global warming, so too will the location of some highly valued resources, sometimes across national boundaries. Are environmental issues different in any way from other international issues? Are existing regional organizations prepared to deal with this "new" source of conflict?

By focusing on two major climate-related issue areas – transboundary water resources and marine fisheries – workshop deliberations could provide useful insights into how well different types of regional organizations have fared in dealing with issues related to regional environmental changes, even in the absence of a global climate change. For example, with regard to the Anglo–Icelandic Cod Wars, the then secretary general of NATO (1957–61), Paul-Henri Spaak, noted in his memoirs the difficulties he encountered in trying to resolve the resource-related conflict between the main protagonists (the UK and Iceland), both of which were key members of the North Atlantic Alliance.[2] This particular conflict led to Icelandic threats to withdraw from use by NATO allies its strategically situated air base at Keflavik. It also led to hostile actions between two NATO members – Iceland and the UK.[3]

---

[2] Spaak, P-H., 1971: *The Continuing Battle: Memoirs of a European 1936–1966*, Weidenfeld and Nicholson, London, UK.

[3] Glantz, M.H., 1992: Global warming impacts on living marine resources: Anglo-Icelandic Cod Wars as an analogy. In *Climate Variability, Climate Change and Fisheries*, ed. M.H. Glantz. Cambridge: Cambridge University Press.

The report of the World Commission on Environment and Development (WCED) reinforced the view that it would be useful to reconsider the utility of regional organizations in coping with the impacts of climate-related environmental change. The report (*Our Common Future*, also known as the Brundtland Report)[4] drew attention to the Brundtland Commission's observation that, "policies formerly considered to be exclusive matters of 'national concern' now have an impact on ecological bases of other nations' development and survival" (p. 311). It recommended that "regional organizations in particular need to do more to integrate environment fully into their macroeconomics, trade, energy, and other sectoral programmes" (pp. 314–315), and noted that existing regional organizations must undertake considerations of environmental issues, calling on them to provide the appropriate level of interest, personnel, and financial support. In fact, a decade earlier the Brandt Commission report[5] stated that "it can no longer be argued that the protection of the environment is an obstacle to development. On the contrary, the case of the natural environment is an essential aspect of development" (p. 114).

The Brundtland Commission also briefly discussed regional organizations from the perspective of sustainable development, noting the strong need to better match international institutions with ecological problems. There is a growing awareness of the fuzziness of distinction between domestic and foreign policy (especially as one country, for example, can plunder its own natural resources, giving little regard to the global implications of its action). The Commission also suggested that intergovernmental regional organizations could gain considerable support from regional non-governmental organizations (NGOs), which can bring different perspectives to roundtable discussions on environmental issues. The WCED briefly highlighted river basins as an example of international ecological systems in need of regional cooperation.

Given that the frequency of such conflicts (or opportunities for cooperation) could increase in the face of global climate change and its regional consequences (e.g., the shifting of highly valued resources),[6] increased attention should be given to the role of regional organizations (both existing as well as those that might yet

---

[4] World Commission on Environment and Development, 1987: *Our Common Future*, Oxford University Press, Oxford, UK.

[5] *North–South: A Programme for Survival*, Report of the Independent Commission on International Development Issues, 1980: The MIT Press, Cambridge, Massachusetts.

[6] T.F. Homer-Dixon, 1991: On the threshold: Environmental changes as causes of acute conflict. *International Security* **16**, 2, 76–116; R. Perelet, 1988: Ecology and diplomacy. *International Affairs* (November), 86–94.

be established) in resource-related regional conflict resolution. It is hoped that this multinational, multidisciplinary NATO Advanced Research Workshop, which involved the participation of resource managers, representatives of regional organizations, political scientists and geographers whose research has focused on regional organizations, decisionmakers who have been (or might be) involved in such resource-related issues, and international law experts interested in international organizations, will have served as a step in that direction.

Key Points

- The decade of the 1990s provides a "window of opportunity" for a review of regional organizations and their potential contribution to the "regionalization of environmental problems."
- Questions about the potential severity at the regional and local levels of impacts of global climate change pertain not only to climatic factors (including extreme events), but also to their relationship to societies possessing different capabilities for implementing proactive as well as reactive policy responses.
- As resources (such as water and fish populations) and people dependent on them "migrate" on land and in the marine environment, the risks of regional conflicts as well as the opportunities for regional cooperation are likely to increase.
- Given that the frequency of such conflicts (or opportunities for cooperation) could increase in the face of global climate change and its regional consequences (e.g., the shifting of highly valued resources), increased attention should be given to the role of regional organizations (both existing as well as those that might yet be established) in resource-related regional conflict resolution.

The Role of Regional Organizations in Addressing Climate Change and Other Complex Environment and Development Issues

Jag S. Maini

Climate change is only one of the several large-scale environment and development issues that has recently appeared on the international scientific, policy, and political agendas. Attempts at addressing these issues requires, first, the formulation of a common understanding of the issue and, second, a commitment to work collectively at the national, regional, and international levels.

While we have millenia of experience within the international community in forming military alliances (e.g., Roman and Ottoman empires, NATO, Warsaw Pact, etc.), and decades or centuries in forming economic and trade alliances (e.g., OECD, GATT, COMECON, ASEAN), our experience in formulating alliances based primarily on environment and development issues is somewhat recent and limited. In order to avoid costly mistakes and delays in the implemention of corrective actions, the scientific and policymaking communities must gain insights from past experiences in coping with other situations of change.

Addressing global-scale issues such as climate change has regional and national dimensions, as well as global ones, in terms of causal factors, consequences, and corrective (i.e., preventive, mitigative, or adaptive) measures. Under what circumstances do countries commit themselves to formulate an objective collective understanding of an issue and cooperate in the formulation and implementation of corrective actions in the interest of the global community? Recent international deliberations that led to the convening of the Earth Summit in Rio de Janeiro, Brazil, in June 1992 provided an interesting experience in the behavior of multinational, regional, and international "tribes." (The Earth Summit was officially referred to as UNCED, the UN Conference on Environment and Development.)

How a regional organization approaches an issue is determined largely by its mandate and its policy framework; for example, possibly along the following lines:

- Political: UN, EEC, ASEAN, G-7, G-77;
- Trade: EEC, NAFTA, Nordic;
- Ecological: The US-Canadian (Great Lakes) International Joint Commission, Amazon Treaty, Antarctica Treaty;
- Commodity: OPEC, International Tropical Timber Organization;
- Legal: Conventions and treaties, such as the Migratory Birds Treaty in North America and Europe, the proposed framework convention on desertification;
- Utilitarian: Rhine and Danube Commissions (dealing mainly with navigation);
- Financial: World Bank, regional banks;

- Industrial: Regional industrial and manufacturing associations;
- Scientific: AAAS, IIASA, ICSU, IPCC;
- Military: NATO.

Many of these regional organizations pursue their particular relatively narrow interests during the various phases of an issue as it evolves, i.e., recognition of early warning signals; denial; elaboration; assessment and definition; search for a common solution; implementation of corrective actions and periodic monitoring of accomplishments. The regional organizations, therefore, become stakeholders that pursue their narrow interests, which may either be in conflict or in convergence with others. For example, many UN agencies, including UNEP, were initially critical and not very supportive of the Brundtland Commission, until the commencement of political deliberations toward the preparation of UNCED. Then they seemed to have come together in support of the concept of "sustainable development" possibly in view of the potential for increased funding. As another example, some OPEC members have refused to recognize that the burning of fossil fuels is directly implicated in the climate change issue and have tried to redirect attention onto other political aspects of global warming such as tropical deforestation and reforestation, and onto other scientific aspects such as the role of the oceans as major sinks for carbon.

It would be useful to examine the possibility of addressing large-scale global issues by formulating regional strategies and action plans. For example, more rapid progress may be made at the regional level in addressing the issue of the health of the oceans and seas. Similarly, the issue of the conservation and sustainable development of global forests may be addressed on a "regional" basis, such as the Amazon Basin, the humid tropics in western Africa, and the circumpolar boreal forests zone. The mobilization of regional organizations may prove to be a critical factor in addressing complex issues related to environment and development such as global climate change. It is hoped that the NATO Workshop discussions will provide some insights on addressing the following concerns:

- How to engage regional organizations constructively, mobilizing their resources and expertise.
- An improved understanding of the institutional dynamics that lead to successes and to failures in multinational regional initiatives.
- What insights might we learn from our hindsight to enhance foresight capability?

Key Points

- Addressing global-scale issues such as climate change has regional and national dimensions, as well as global ones, in terms of causal factors, consequences, and corrective measures (e.g., preventive, mitigative, or adaptive).
- How a regional organization approaches an issue is determined largely by its mandate and its policy framework.
- It would be useful to examine the possibility of addressing large-scale global issues by formulating regional strategies and action plans. For example, more rapid progress may be made at the regional level in addressing the issue of the health of the oceans and seas. Similarly, the issue of the conservation and sustainable development of global forests may be addressed on a "regional" basis.
- The mobilization of regional organizations may prove to be a critical factor in addressing complex issues related to environment and development such as global climate change.

The Climate Change Issue: Scientific Aspects

Robert S. Kandel

Research on climate involves scientists from many disciplines, and that support for such research has increased in recent years. Media exposure of vocal scientists involved in the climate change issue has also grown and, as a result, scientists may have a greater opportunity to influence societal decisions than they had in the past. The media, like many scientists, tend to given inadequate attention to the uncertainties that surround the scientific aspects of of global warming. As a result of the media's search for extreme (i.e., spectacular) views on global warming, many scientists have become reluctant to share their research results with the media.

Climate may be taken as defining the physical conditions of existence of life at or near the surface of the Earth. Different geographical regions have different climates. Leaving aside the nightmare of a global catastrophe eradicating all life on our planet, examination of the ecological and societal impacts of climate change must focus on the effects of climate changes at the regional level.

Climate on the regional scale is governed by various aspects of the global climate system. This system is constantly changing; the critical issue concerns the degree of disequilibrium in the system. The reflection and emission of energy to space depend on many natural and anthropogenic factors. The effects of variations deriving from natural processes, such as volcanic eruptions, on the atmospheric system tend to be relatively short-lived (on the order of a few years). In contrast, many human activities can result in long-term cumulative effects on the concentrations of atmospheric chemicals and particulates and on the amount and types of cloud cover.

The global climate system operates by converting into heat about 70% of the energy flux associated with visible and near infrared solar radiation. This heat is ultimately re-radiated to space in the form of medium infrared radiation in the 4-40 μm wavelength range. It is important to note that within this range, the system operates a geographical redistribution of energy by way of the general circulation of the atmosphere and oceans. This redistribution is clearly coupled to the hydrologic cycle which brings water to the continental surfaces. Concern about human-activities-induced global warming arises because: (i) surface temperatures depend in part on the opacity of the atmosphere to the infrared radiation which ultimately must be emitted to space; and (ii) infrared opacity depends on the atmospheric concentrations of polyatomic molecules ("greenhouse gases"), some of which (carbon dioxide, methane, nitrous oxides, the chlorofluorocarbons) are increasing primarily as a result of human industrial and agricultural activities.

The activities which produce greenhouse gases vary greatly from one region to the next. However, because of the rapidity of the atmospheric circulation and the relatively long atmospheric residence times of these gases, greenhouse gas enrichment is global. Continuation of current trends will certainly reinforce the naturally occurring greenhouse effect and would lead to some degree of global warming over the next several decades. Despite reports on consensus, however, there remains considerable disagreement on such important issues as the following: (i) the likely course of changes in greenhouse gas concentrations during this period, and how different policies might affect it; (ii) the sensitivity of the climate system to a given change in greenhouse gas abundance. This sensitivity is often expressed as a global mean equilibrium temperature change for a total greenhouse "forcing" (that is, a trapping of infrared radiation flux) of 4 W/m^2. This is "equivalent" to the effect of a carbon dioxide doubling alone.

Model estimates of the sensitivity per 4 W/m^2 forcing range from 1 to 5°K. Although there are some dissenting views, all of the modelers' estimates are based on assumptions about the "water-vapor feedback," that is, they assume that warming will necessarily be accompanied by an intensification of the hydrologic cycle and an increase in atmospheric water vapor content, positively reinforcing the greenhouse effect. Disagreements arise mainly over the role attributed to changes in cloud cover which will accompany changes in the hydrologic cycle, because the likely changes in amounts and types of clouds are still unknown. Although modern climate models are more or less successful in simulating observed features of the geographical distribution of regional climates and their seasonal cycles, they remain inadequate in many respects for a variety of reasons (e.g., grid size is too large and obscures regional detail). Thus, little confidence can yet be given to model projections of regional climate changes that might accompany global warming. This becomes particularly obvious when comparing maps of different model projections of changes in temperature. They are in even less agreement in their projections of precipitation and soil moisture, that is, in the distribution of regional water resources.

The extreme uncertainty of model estimates of regional climate changes expected to accompany a global warming means that it will be difficult to identify such a trend with confidence, as long as the temperature and precipitation changes are small. In other words, the CO_2 "fingerprint" that scientists and policymakers alike hope to identify remains "blurred." If there is no convincing progress in model estimates, it will be difficult to estimate climate sensitivity before changes become significant. By then, it will probably be harder to reduce greenhouse gas emissions at a pace fast enough to avoid still larger, unwanted climate changes. The uncertainty that surrounds model estimates also means that, in seeking to develop societal adaptation strategies to, rather than prevention of, climate change, planners can only consider the regional climate change estimates as plausible but not necessarily reliable scenarios of future change. Any particular scenario may have quite different implications for different regions; and different plausible scenarios may have very different implications for a given region.

Key Points

- Examination of the ecological and societal impacts of climate change must focus on the effects of climate changes at the regional level.

- The activities which produce greenhouse gases are greatly differentiated from one region to the next. However, because of the rapidity of the atmospheric circulation and the relatively long atmospheric residence times of these gases, greenhouse gas enrichment is global.

- Despite reports on consensus, there remains considerable disagreement on such important issues as (i) the likely course of changes in greenhouse gas concentrations during this period, and how different policies might affect it; and (ii) the sensitivity of the climate system, in particular as moderated or amplified by water vapor and cloud feedbacks, to changes in greenhouse gas abundances.

- Although modern climate models are more or less successful in simulating observed features of the geographical distribution of regional climates and their seasonal cycles, they disagree in many aspects of their predictions of regional climate changes, and remain inadequate in their treatment of ocean–atmosphere coupling and of the water cycle. As a result, little confidence can yet be given to model projections of regional climate changes that might accompany global warming.

- Planners can only consider the regional climate change estimates as plausible but not necessarily reliable scenarios of future change. Any particular scenario may have quite different implications for different regions; and different plausible scenarios may have very different implications for a given region.

The Climate Change Issue: Policy Aspects

James P. Bruce

The policy perspectives on the global warming issue can be characterized as being similar to such issues as acid precipitation and stratospheric ozone depletion, as a problem of "contamination" of the global atmosphere by human activities. At the request of the UN General Assembly, the World Meteorological Organization (WMO) and the UN Environment Programme (UNEP) established the Intergovernmental Panel on Climate Change (IPCC) in 1988 for the purpose of undertaking scientific assessments of global warming and its implications for ecosystems, society, and policy. A key element in such an assessment is an inventory of the major terrestrial, oceanic, and anthropogenic sources and sinks of greenhouse gases (GHGs) which, at present, are not very well known.

Major anthropogenic sources of GHGs include but are not limited to the combustion of fossil fuels, biomass burning, and land-use changes (e.g., tropical deforestation). According to the IPCC's "business as usual" scenario, rates of CO_2 emissions from developing countries are expected to continue to increase faster than those from industrialized countries. This implies that, if the atmospheric concentration of CO_2 is to be stabilized at some point in the future, in addition to industrialized countries decreasing their dependence on fossil fuels, future paths of development (involving technology transfer) in other countries should be designed to minimize CO_2 and other greenhouse gas emissions. The eventual atmospheric concentrations of various greenhouse gases will depend on the choice of development strategies in all parts of the globe and the time period over which they are implemented.

Recent studies suggest that the implementation of many already existing technologies which decrease CO_2 emissions is cost-effective and could in most countries decrease by 10–20% both energy use and CO_2 emissions, in addition to a decrease of emissions of other pollutants. The implementation of such energy-efficient technologies, however, requires some structural changes, such as a decrease in subsidies that encourage fossil energy use, an increase in the use of alternative energy sources (e.g., wind, biomass), and the production of other energy sources (e.g., solar–thermal, photovoltaic) that could soon become competitive.

The Framework Convention on Climate Change, signed by 154 countries at the UN Conference on Environment and Development (UNCED) in Rio de Janeiro in June 1992, is now in the process of ratification by national governments. Once ratified, 36 of the signatory states are expected to file plans to limit GHG emissions within one year of signing. The most important initial action will be to reduce fossil fuel use and to increase energy efficiency. Despite such actions, some degree of global warming can be expected to occur because of past human activities (some GHGs have a long residence time in the atmosphere) and the time necessary to implement new controls on greenhouse gas emissions. Thus, adaptive measures designed to cope with regional changes of global warming are essential for climate-sensitive socioeconomic sectors such as agriculture, water resources and, particularly, natural hazards. Global warming is likely to result in an increase in the geographical range and perhaps intensity of extreme meteorological events, such as storm surges, floods, droughts, and tropical cyclones. Many technological and institutional strategies for decreasing human and economic losses from such events are already available, and should be implemented.

Regional organizations come in a bewildering array of sizes and functions, and their roles in the climate change issue are equally diverse. It may be useful to consider their roles within the framework of the following activities associated with climate change:

data collection, climate system research, impact assessments, adaptation strategies, and control programs.

Data collection

Measurements of components, trends and variability of the climate system, the atmosphere and climatic parameters, oceans, cryosphere and biosphere should be carried out based on globally accepted procedures and methods. The implementation of the measurement programs is primarily a national responsibility but is often reinforced and made more effective through regional action. Standard observational procedures are usually established through UN organizations. The regional organizations involved are mostly regional groupings within the UN organizations, e.g., WMO's six Regional Associations, UNEP's Regional Seas Program organizations, and offshoots such as SPREP (the South Pacific Regional Environment Program).

Climate system research

Regional research efforts within the framework of the World Climate Research Program and the International Geosphere Biosphere Program have been organized somewhat differently. For example, key projects under the Tropical Oceans Global Atmosphere (TOGA) project to develop the capability of predicting ENSO (El Niño–Southern Oscillation) conditions, have focused on the Equatorial Pacific. In the Global Energy and Water Cycle Experiment (GEWEX) program, regional studies of large river basins, such as the Mississippi (USA-Canada), the Mackenzie (Canada) are a key element.

The START (System for Analysis Research and Training) initiative to organize a number of regional research networks, especially in developing regions, shows enormous promise. An Inter-American Institute, a network of centers for global change research, has been agreed to by 20 or so countries of the Americas. The UN's Global Environmental Facility will be funding a network for the ASEAN countries with a center in Indonesia and the main meteorological and climatological work based at the Regional Meteorological Centre (WMO) in Singapore. Development of a strategy for Africa is under way, and will likely build upon the work of the continent-wide ACMAD (African Centre of Meteorological Application for Development), the WMO African Region, the Economic Commission for Africa, UNEP-funded climate activities throughout the continent, and the AGRHYMET (Agriculture–Hydrology–Meteorology)

Centre (WMO/FAO/CILSS) for regional studies and applications of meteorological and hydrological information to improve agricultural production in the Sahel.

Impact assessment

A great deal of work can be done through a variety of regional organizations on assessing the impacts of climate variability and change. The following examples illustrate the range of possibilities and are suggestive.

In the North American Great Lakes Basin, initial studies suggest drastic lowering of lake levels and flows and a "second generation" impact study is under way through Canada–USA regional arrangements. However, the Canada–USA International Joint Commission is undertaking a study of possible further regulation of Great Lakes levels, and the climate change issue, as well as increasing consumptive use, are becoming important aspects.

The regional Asian Development Bank is funding a set of studies of climate impact and response strategies for eight countries (Bangladesh, India, Indonesia, Malaysia, Pakistan, Philippines, Sri Lanka and Vietnam).

The UNEP-supported World Climate Impact and Response Strategies Program (WCIRP), a component of the World Climate Program, has given its highest priority to regional impact studies, including most recently those under way in southern Africa, Southeast Asia, and eastern Europe and on coastal zone impacts related to the Regional Seas Action Plans.

Adaptation strategies

Adaptation to the consequences of global warming, including sea level rise, is an important priority in many countries, and in some areas a regional approach can be helpful in devising appropriate response strategies. This is especially true, for example, in connection with resolution of water resource conflicts, sea level rise, and reduction of losses due to natural disasters.

Resolution of conflicts related to water apportionment, transboundary water pollution and flood warnings will likely become increasingly important, with changes in evaporation and precipitation regimes in the 47% of the world's land surface that lies within international drainage basins. In many regions significant regional conflicts are developing. A number of types of international basin mechanisms have already been employed, with varying degrees of success for recent water-resource-related conflicts, but in many international basins no such mechanisms exist.

Declaring in the early 1990s the International Decade for Natural Disaster Reduction (IDNDR) has been a useful way to foster regional cooperation. Such cooperation has been promoted effectively by such regional organizations as the Pan-American Health Organization (PAHO) of WHO, and the Organization of American States (OAS) for Latin America and the Caribbean, the Organization of African Unity (OAU) for that continent, and the Asian Disaster Preparedness Centre and Asian Development Bank for Asia and the southwest Pacific region.

Control strategies

Control of the rate of climate change would require placing limits on greenhouse gas emissions in all parts of the world. Because of the intense economic and political sensitivity and the disparities among countries in their sources of emissions of carbon dioxide, methane, and other radiatively active trace gases, studies of the feasibility of emission reductions and action programs are being undertaken at the national level. However, countries within a region can learn from each other about the benefits, costs and feasibility of various strategies. To this end, a number of regional organizations have proven effective in facilitating exchanges of information and perspectives, such as the OECD and UN Economic Commission for Europe. It would be useful to encourage greater efforts by other UN regional economic commissions to foster greenhouse gas limitation strategies within their regions as part of an overall effort at sustainable development planning at the regional level.

Key Points

- In addition to industrialized countries decreasing their dependence on fossil fuels, future paths of development (involving technology transfer) in other countries should be designed to minimize CO_2 and other greenhouse gas emissions.

- As some degree of global warming can be expected to occur because of past human activities, adaptive measures designed to cope with regional changes of global warming are essential for climate-sensitive sectors such as agriculture, water resources, and natural hazards.

- Regional organizations come in a bewildering array of sizes and functions, and their roles in the climate change issue are equally diverse. It may be useful to consider their roles within the framework of activities associated with climate change: data collection, climate system research, impact assessments, adaptation strategies, and control programs.

- In some areas a regional approach can be helpful in devising appropriate response strategies. This is especially true, for example, in connection with resolution of water resource conflicts, sea level rise, and reduction of losses due to natural disasters.

- Because of the intense economic and political sensitivity and the disparities among countries in their sources of emissions of GHGs, countries within a region can learn from each other about the benefits, costs and feasibility of various emission reduction and action strategies.

Transnational Regional Responses to Global Climate Change: Options – Obstacles – Opportunities

Lynton K. Caldwell

The prospect of significant changes in the climate of the Earth raises questions regarding resulting changes in ecology, economics, populations, and politics. As best we can foresee, these changes are likely to be both global (e.g., sea level rise) and regional (e.g., change in rainfall, temperature, and extreme events). The social, political, and ideational characteristics of today's world are poorly adapted to responding to major geobiospheric changes. Adaptive responses are unpredictable under present uncertainties. Yet, a range of possible responses may be conjectured on the basis of present estimates and past experiences with transnational regional responses to perceived common threats (e.g., UNEP's Regional Seas Programme). We may identify options for new transboundary arrangements and the opportunities that alternative choices present. But a realistic approach to these conjectures requires the identification of the obstacles to regional and national cooperation and the devising of credible alternatives to institutional inadequacies.

Regional organizations may be categorized in many different ways. In general, they include representatives from three or more states, with some notable exceptions, such as in North America where binational organizations (between Canada and the USA, and between USA and Mexico) operate in large, relatively well-defined regions. For any issue, a region must be defined by certain common denominators, which can be used in assessing whether existing organizations have, or can develop, the capability to act. To date, few regional organizations have become concerned with the climate change issue. It is difficult to define regions in the context of global warming, although enclosed seas might be identified as one type of coherent, directly concerned region.

Regional organizations are created by states because they find it increasingly difficult to deal effectively with numerous international issues of varying levels of national importance that require constant attention. Regional organizations have four advantages:

1) *Flexibility of collective action.* If the goals of collective action are agreed to, diplomatic interactions are less likely to encumber procedures, and new required expertise more easily identified or developed. However, scientific and diplomatic goals and actions can often be in conflict.

2) *Capability to initiate policy.* The regional organization can take responsibility for policies which individual governments might not wish or be able to explicitly propose.

3) *Negotiation.* Representatives of member states can interact in a different forum than that required for direct diplomatic channels. Goals in common have already been accepted.

4) *Exchange of views.* Governments, non-governmental organizations (NGOs), and industries are more likely to provide inputs to, as well as to influence, multilateral policy development.

The historical record demonstrates that the NEED for basic changes in social institutions characteristically emerges well in advance of political recognition of the NECESSITY for such change. When a problem pointing toward a need is first perceived by a far-sighted few, the majority response tends, at first, to be one of dismissal. As the evidence of a problem grows, the response often turns to denial. Initially, the problem is itself denied, particularly by persons (e.g., public officials) who would be responsible for addressing it. But if the evidence of danger ahead becomes too compelling to dismiss, denial of the existence of the threat turns to denial that anything can be done about it. People are advised to adapt to changing conditions on the argument that there is insufficient evidence that the alleged problem is a reality. And if it is indeed real, then the problem is unmanageable, since the cost of remedial action would wreck the economy, frustrate human aspirations and – worst of all from an official perspective

– turn people against the government. Moreover, the task of obtaining the necessary commitment of all governments concerned would be daunting.

Options

The nations of the world recently concluded the largest United Nations conference ever (UNCED) with the adoption of the massive detailed Agenda 21 to guide national and international action in the 21st Century. Although the major environmental problems addressed were global in scope, they all had regional implications. Climate change was the largest issue, since the depletion of the ozone layer had already been addressed and, in large measure, resolved politically before the convening of UNCED in June 1992.

The depletion of stratospheric ozone and the prospect of global climate change have both been induced by controllable (in theory) human actions. The basic differences between these two atmospheric problems, however, have led to distinctly different policy scenarios. For example, the ozone issue was relatively more simple and more immediate. Its causes and progression could be measured and its principal dangers could be foreseen with greater confidence than those stemming from global climate change. Moreover, remedial measures for ozone depletion could be taken with relatively little impact upon economies and lifestyles. The impact of ozone depletion was directly personal: risks to health, out-of-doors activities, and food production. Scientific consensus on the stratospheric ozone depletion issue was sufficient enough to move governments to action. The 1985 Vienna Convention, the 1987 Montreal Protocol, and various subsequent meetings and declarations left the decisionmakers with few, if any, options. Even the conservative, laissez-faire government of the United States hastened to ban chlorofluorocarbons (CFCs) as early as the mid-1970s. Aside from relative simplicity, the option to ban CFCs was aided by the easily identified localized manufacture of CFCs and the fact that the regionalization of the environmental threat first occurred over unpopulated Antarctica and the Southern Ocean. No elaborate international institutional machinery seemed necessary to implement the ozone agreements.

The circumstances that surround the global climate change issue are, in fact, very different. The causes of changes in atmospheric chemistry related to global warming are more complex and less clearly understood. Remedial measures and adaptive responses implied far-reaching, costly, and controversial changes in economic, technological, and political behavior. Moreover, these changes would necessarily be multinational and would need to provide for the different consequences of change that might be expected to occur among the Earth's regions. Preventative and remedial action would severely impact energy production from petroleum and coal and methane production in agriculture. It would reverse the process of deforestation. Although rapid by geological standards, in human terms global climate change as presently projected appears to be slow and, for the present, unobtrusive, since noticeable effects are not expected to occur for several more decades. It is not in the nature of practical politics to give much attention to circumstances that are not presently compelling or that are not likely to arouse public concern in the near future. In a world of many troubles, politicians are reluctant to ask for present-day sacrifices on behalf of a future (an uncertain one) that most of their constituents will not live to see.

The options for dealing with prospective global climate change or its potential regional consequences are not self-evident. The technical and behavioral changes that could arrest global warming are known, and many would merit adoption even if global warming were not an issue. For example, reduction of carbon dioxide and other "greenhouse" gas emissions could have collateral benefits for promoting research in energy production, in agriculture, and in the use and protection of forests. Energy dependence on fossil fuels is on "borrowed time." Nevertheless, the energy needs of human society involve a large number of issues that most governments prefer to dismiss or deny. Principal among these is the issue of population growth, distribution, and per-capita demands. Evasion of basic factors, such as population demand and the ultimate decline in resource availability, is regarded by most governments as a viable political strategy. And because of uncertainties regarding the magnitude, timing, and incidence of global climate change, delaying and avoiding "precipitous action" have been attractive options for many policymakers, most notably those in the United States.

Obstacles

An obvious "obstacle" to a considered and effective response to the prospects of a global climate change and its regional impacts is the "wait and see" option – to postpone decisionmaking to a future time when the need for action will likely be more compelling. This reluctance to "borrow trouble" from the future is one aspect of a larger phenomenon – "institutional lag." The lag syndrome is the sum of a number of different responses to threatening developments in society that might imply a need for new institutional arrangements.

A common reaction, as has been stated, is the initial denial that a threat exists. A threat may subsequently be conceded, but skeptics may regard it as relatively minor. Existing institutional arrangements are considered to be adequate to deal with it, especially if alternatives would require changes in the law (notably in jurisdiction) and in the structure of public policy and administration. Public bureaucracies are especially protective of their prerogatives. National agencies are seldom ready to subordinate their autonomy to transnational superordinate authority.

Underlying and reinforcing this introversive tendency in government agencies (a tendency not confined to government) is the political doctrine of sovereignty. This doctrine, which originally defined the prerogatives of rulers, was extended to the modern state. Acknowledgment of the inviolability of national sovereignty occurs in the preamble and elsewhere in the declarations, proposals, and action plans adopted by United Nations conferences; and the 1992 Earth Summit (UNCED) was no exception.

A more rational obstacle to national responses to transnational problems is an honest uncertainty as to the most feasible and effective way to address or resolve those problems. The obvious inclination of most politicians and public officials is to adapt existing national institutions to the problem. When this option seems insufficient, the usual inclination is to proceed with caution, thus creating transboundary or regional bodies with strictly delimited powers and provisions for national noncompliance on disputed propositions. The International Whaling Commission offers a textbook example of national "reserved" consent (the right of a nation to reject a resolution as contrary to its interests).

An obstacle to international joint effort that appears to be diminishing has been the absence of a politically significant transboundary popular constituency. International agreements and even institutions have been established at the behest of particular interests, chiefly economic, whose members form international networks and have the collective power to move governments. GATT, IMF, the European Community, and the projected North American Free Trade Association (NAFTA) are examples of such international and regional trade and monetary agreements. The earlier regional commissions dealing with environment-related issues (e.g., the Rhine, Danube, and the Great Lakes) were primarily concerned with navigational and jurisdictional issues. International public health agreements relating to regionally localized diseases were advanced by a broader popular concern over the possibility of wider contagion.

The emergence of an international environmental movement and network of NGOs has begun to remove the obstacle of an insufficiently organized transnational constituency able to command the attention of governments. It was hardly a sudden conversion to "environmentalism" that brought more than one hundred heads of state to the UN Conference on Environment and Development. Political forces are now working internationally that may cause the commitments of the UNCED Agenda 21 report to have a decisive effect upon world affairs in the future.

Nationalism as a popular civic religion is diminishing in techno-economically advanced countries. But the problem of consensus becomes more difficult to resolve in pluralistic republics with diverse organized interests. The rise of international linkages among NGOs may improve the prospects for regional and international cooperation on environmental issues.

Opportunities

The pace of history – of multiple changes throughout the world – appears to be accelerating. Institutional lag will handicap efforts to deal with an array of environmental and natural resource problems that must be addressed in the interest of human welfare if the future is to be ecologically and economically sustainable. But institutions may outlive their usefulness and continue to be obstructive unless credible alternatives are available. Some analysts (e.g., Donella Meadows, *Beyond and the Limits*, 1992) see institutional and economic collapse as real dangers. Avoidance of climate-related environmental disaster appears possible, but it would require timely action which governments and many people are unprepared or unwilling to take.

If modern society is to avoid being caught unprepared to cope with environmental changes that may arrive earlier than expected, exploratory and conjectural studies of alternative futures will be needed. We do have some experience with institutional innovation responsive to transnational problems. In Europe, the Mediterranean Blue Plan and cooperative actions in the Baltic Sea, the North Sea and the wider Caribbean provide at least partial models of the conditions and courses of responsive institutional development. In North America, the International Joint Commission has evolved in response to public concern over the deterioration of the Great Lakes ecosystem. In nearly every case, however, goal attainment by regional organizations is thwarted or delayed because of "foot-dragging" tactics by governments and their "turf conscious" agencies. Foreign affairs offices are notoriously cautionary and legalistic, and characteristically the natural resources and environmental protection agencies jealously guard their mandates and bureaucratic interests.

Global warming is one of several complex environmental issues that is leading to an increase and diversity of mechanisms for influencing policy at the national and international levels. This derives both from the fact that communication and meetings are more open, as well as from the difficulties that governments are experiencing in dealing with several simultaneously occurring unprecedented complex environmental, social, economic and political problems. In the post-UNCED period, NGOs have been playing an increasing role in national and international policy development. This trend toward greater NGO participation is associated (some argue) with an apparent decrease in the sovereignty of governments (e.g., the adoption of sustainable environmental policies as conditions for technical or financial assistance. Nevertheless, sovereign governments still make the ultimate decisions, such as signing and ratifying treaties.

In conclusion, an opportunity exists for a productive linkage between research, forecasting, public participation, and institutional innovation. How far the opportunity can or will be taken remains uncertain. Organizing and funding such linkages will require changing the priorities hitherto prevailing in governments and international organizations, in philanthropic foundations, and in research institutions (especially in universities).

Key Points

- We may identify options for new transboundary arrangements and the opportunities that alternative choices present. But a realistic approach to these conjectures requires the identification of the obstacles to regional and national cooperation and the devising of credible alternatives to institutional inadequacies.
- Regional organizations are created by states because they find it increasingly difficult to deal effectively with numerous international issues of varying levels of national importance that require constant attention. Such organizations have at least four advantages: Flexibility of collective action, capability to initiate policy, negotiation, and exchange of views.
- The historical record demonstrates that the NEED for basic changes in social institutions characteristically emerges well in advance of political recognition of the NECESSITY for such change.

- This reluctance to "borrow trouble" from the future is one aspect of a larger phenomenon – "institutional lag." The lag syndrome is the sum of a number of evasive responses to threatening developments in society that might imply a need for new institutional arrangements.

- National agencies are seldom ready to subordinate their autonomy to transnational superordinate authority. In nearly every case, goal attainment by regional organizations is thwarted or delayed because of "foot-dragging" tactics by governments and their "turf conscious" agencies.

- The emergence of an international environmental movement and network of NGOs has begun to remove the obstacle of an insufficiently organized transnational constituency able to command the attention of governments. It was hardly a sudden conversion to "environmentalism" that brought more than one hundred heads of state to UNCED.

An Environmental Security Dimension of Global Climate Change

Renat Perelet

Two major converging trends have recently appeared: (1) the notion of security has acquired a specific orientation such as food security, humanitarian security, economic security, and environmental security; and (2) natural resources, including climate, are now considered to be part of the broader notion of the environment. The resulting uneven local and regional impacts of global environmental changes such as climate variability on countries and people or on their access to environmental resources has increasingly contributed to tensions within states as well as between them. Furthermore, the fact that the property rights to what have been called common property resources (e.g., the atmosphere, outer space, the high seas, etc.) remain unclear, exacerbates interstate tension.

Aside from the impacts of natural changes in the environment, governments can intentionally use the environment as a weapon (i.e., environmental warfare). The intentional use of the environment by a state or region as a weapon in settling disputes with other states or regions was most recently observed during the Gulf War, when retreating Iraqi troops set hundreds of Kuwaiti oil wells on fire. Because environmental changes may become a cause of acute conflict, the environment has become a legitimate security issue.[1] Thus, any redefinition of security must include, in particular, an

[1] T. Homer-Dixon, 1991: On the threshold: Environmental changes as causes of acute conflict, *International Security*, Vol. 16, No. 2, pp. 76–116.

appreciation of the importance of resources and the environment as key components of national security.[2]

It is not the environment per se that serves as the main cause of tension within or between countries. It is often the human interactions with and pressures on the physical environment that lead to conflicts. In addition to the scarcity of traditional natural resources (such as fossil fuels), attention has increasingly focused on additional scarcities such as potable water supplies, fertile topsoil, and conditionally renewable natural resources (e.g., forests, biodiversity). Uncertainties that surround our knowledge of natural environmental processes further complicate the situation. Four major factors that affect environment-related regional or global security are as follows: heterogeneous perceptions of global environmental change (GEC) by countries, the uneven impact of GEC on different countries, differing contributions of countries to causes of GEC, a heterogeneous national or regional capacity to respond to adverse GEC. Environmental issues are often wittingly tied in to other issues and used as bargaining chips in international negotiations. As a result of this tie-in tactic, and because GEC will probably raise the level of stress in the international system, increasing the likelihood of conflict, the international dimension of GEC has recently received growing attention. It may also force us to revise our traditional definitions of such concepts as "state," "sovereignty," "national interest" and "balance of power."[3]

The distinction between threats and vulnerabilities points to a key division in security policy, namely that states can seek to reduce their insecurity either by reducing their vulnerability or by preventing or lessening perceived (as well as real) threats. Global policy problems involve relationships between states that call for the management of areas beyond their political jurisdiction. Regional environmental threats often result from transboundary environmental problems: for example, the waters around Denmark are polluted by other North and Baltic Sea countries. The Chernobyl accident serves as yet another example in a growing list of international environmental threats that stem from activities within the borders of one country. Tropical deforestion in the Amazon and pollution of Lake Baikal are further examples. In addition to inadvertent "environmental aggression" through the transboundary transfer of air and water pollutants, international trade has become a transboundary carrier of environmentally unsound technologies, industries, goods, and wastes.

---

[2] S. Lonergan, 1992: Redefining security. *Delta*, Vol. 3, No. 2, Fall, p. 8.

[3] T. Homer-Dixon, 1990: Environmental change and security. *Delta*, Vol. 1, No. 2, Winter, p. 3.

Transborder human migration is another issue that impinges on environmental security. Environmental changes are converting what might have normally been severe food shortages into major famines. These, in turn, produce growing streams of refugees. UNEP has estimated that by the year 2000, one billion environmental refugees will have been displaced from their homelands because their basic means of survival will have been destroyed.[4]

The perception of the widening disparities in environmental quality among countries (whether or not those perceptions accurately reflect reality) enable governments to seek foreign culprits to blame. The option to internalize trade in eco-externalities becomes an important tactic to settle environment-related disputes. For example, it has been argued that the entire Middle East faces increasingly grave and tangled problems of water scarcity that can affect the region's "stability." As another example, it has been estimated that tens of millions of Chinese are trying to leave their country's impoverished interior and northern regions – where water and fuel wood are desperately scarce and the land often badly damaged – for China's economically booming coastal cities. This migration could lead to bitter disputes among these regions over problems related to the consequences of water-sharing and shifting populations.[5]

Causes of environmental insufficiency and disparity

National environmental security (and sovereignty) can be affected by at least three major factors:

- transboundary, regional, or global environmental change that has (or is perceived to have) adverse consequences for a nation's environment. The resulting depletion or adverse changes in the quality of environmental resources causes countries to seek to gain ccess to those resources outside their border;
- international trade relationships that exploit one country's environmental resources (including air and water quality). This includes the transfer of environmentally unsound technologies, industries, and waste. A disadvantaged trading party can become a source of environmental "goods" to some countries, as well as a sink for environmental "bads" to others; and

[4] F. Nuscheler, 1991: Refugees flooding islands of affluence. *Development Forum*, Vol. 19, No. 3, p. 19.

[5] T. Homer-Dixon, J. Boutwell, and G. Rathjens, 1993: Environmental change and violent conflict, *Scientific American*, Vol. 268, No. 2, pp. 38–45.

- environmentally unsound economic development strategies within a country.

Adverse environmental threats to or effects on other states can be accidental or planned, isolated or cumulative, natural or human-induced. International responsibility (transboundary, regional and global) will grow in importance with regard to the settlement of environmental disputes.

The uneven distribution among countries of environment-related resources causes domestic as well as international tensions. There should be an identifiable minimal level of regional, if not national, sufficiency with respect to the environmental resources to avert intra- and interstate conflicts. In addition, extreme environmental resource differences between nations or regions should be somewhat mitigated to achieve some degree of international environmental resource parity. (Parallels are drawn here to military, humanitarian as well as social security issues.) International action (perhaps under the auspices of the UN) could be required to meet the above two conditions, which are necessary for achieving sustainable development and environmental security.

Perhaps a UN document similar to the UN Declaration on Human Rights would be needed in the environmental area, but one that would be operational and enforceable, one that would be formalized in order to address the disparities between different countries.

A change is needed from emphasizing socioeconomic development to a socio-ecological development featuring environmental protection and sustainable development with an effective economic mechanism for allocating resources to meet basic human wants and needs. Nowadays, environmental issues figure prominently on the North–South agenda. This requires focus on the following:

- an international security aspect of sustainable development, including the notions of environmental parity and environmental sufficiency (definitions, criteria indicators, assessment techniques, realization and control mechanisms);
- an analysis of the existing mechanism(s) of international environmental management;
- a modeling basis for a global and/or regional environmental system.

The dramatic political changes in the former Soviet Union and Eastern Europe since the mid-1980s, along with the end of the Cold War, have apparently led to a shift from the dominant international concern about the possibility of a global thermonuclear war to a concern about scores of local and regional ethnic conflicts, with a high risk for eventual international community engagement in their resolutions.

Redefining security

A decade ago Johann Galtung pointed out that security is based on the goals of the environment and development systems.[6] At the same time, the Palmer Commission advocated "common security"[7] focusing on the mutual vulnerability of contemporary societies that challenges the usefulness of unilateral national military security and calls for mutual cooperation and disarmament. The Commission, however, did not consider the environmental dimension.

Recently, Lodgaard stressed that the traditional goal of national security policies had been twofold: (1) to preserve territorial integrity and (2) to preserve the right of national self-determination. He viewed the concept of environmental security as fairly precise and politically useful, although still rather controversial. He emphasized that the resolution of the growing number of environment-related conflicts – humans against nature (a question of sustainability), and humans against other humans (a question of development) – has become more urgent. He suggested that the key to such conflict resolution is sustainable development.[8]

Security at the national level is provided by the state to its people. It has never been limited to military security but has been just one of several security-maintaining instruments, covering traditional social (including human rights), medical, and economic activities (protecting entrepreneurial and other economic activities). The international dimension of national security includes the creation and maintenance of an international climate (in a sociopolitical sense) conducive to a country's sustainable development objectives.

The impossibility for each nation to perform these tasks alone has led them to create military, political, and economic alliances centered on some aspect of mutual interest (e.g., territorial, ideological, cultural, or humanitarian grounds). In such arrangements,

---

[6] J. Galtung, 1982: *Environmental Development and Military Activity: Towards Alternative Security Doctrines*, Oslo, Universitetsforkaget.

[7] Independent Commission on Disarmament and Security Issues, 1982: London: Pan.

[8] S. Lodgaard, 1992: Competing schemes for Europe: The CSCE, NATO and the European Union, *Security Dialogue*, Vol. 23, No. 3, September, pp. 65-66.

states always give away some part of their decisionmaking capacity to an international entity (through membership to a treaty, convention, agreement, organization).

Environmental security issues challenge the political organization of "development" and the perpetuation of modern (state) models of social organization. It also challenges the global political order to establish new innovative modes of governance. For example, environmental movements are also operating in ways that challenge the notion of the state as analogous to a nonporous political container.[9] So too are transnational corpirations. In many ways, environment-related issues highlight the inadequacy of market- and state-run economic arrangements to meet the basic needs of populations. Security cannot be provided in the long term by assuming that the miracles of the marketplace or military superiority will always provide the answer.[10]

Environmental themes have already facilitated the construction of new political networks, linking government agencies, nongovernmental organizations, and grassroots environmental groups into an expanding global civil society. In linking environment solely to state-defined understanding of security, there is a danger of only addressing government agencies and advocating their survival.[11] In fact, in the mid-1980s, three-quarters of the world's development was financed by the private sector.[12] In addition to governments, the most important political actors are often local citizens' groups, social movements, global and regional nongovernmental environmental organizations, and transnational corporations, among others.

As a final comment, it is sometimes argued that linking environmental themes to national security is a mistake which leads governments away from dealing adequately with the environmental challenges already facing the planet. However, in this view security is narrowly defined as being handled primarily by military means which, in fact, are often inappropriate against environmental threats. As Brock recently noted, "Defining environmental issues in terms of security risks is in itself a risky operation.

---

[9] S. Dalby, 1992: Ecopolitical discourse: Environmental security and political geography, *Progress in Human Geography*, Vol. 16, no. 4, p. 504.

[10] Dalby, p. 518.

[11] Dalby, p. 517.

[12] UNEP (United Nations Environment Programme) Press, 1985: 14 May 1985, 85/9, p. 7.

... We may end up contributing more to the militarization of environmental politics than to the demilitarization of security politics."[13]

Moss, too, questioned the usefulness of the notion of environmental security. He noted that the notion could be used to call attention to the fact that "the security of the environment" needs to be protected from the threats posed to it by human activities.[14] It could also be used to imply that threats posed by global change or environmental degradation could have something in common with the sorts of threats of organized violence that are normally considered threatening to national security.

Moss gives preference to the first meaning. In this interpretation, the achievement of environmental security is linked with environmentally safe human (or technological) activity or, as he rightly puts it, it is equated with "sustainable development." This is different from the extension of the notion of security by simply adding an environmental dimension. He argued that because threats to "security" typically emanate from outside a state's borders, viewing environmental problems as threats to security is likely to direct attention away from one's own contributions to transboundary environmental problems.

The concept of security has evolved into a view that embraces the interlocking elements of environmental security, individual security, societal security, economic security, and military security.[15] It is now abundantly clear that the insecurities that first occur in regions facing serious environmental problems, particularly in the Third World, spread quickly to engulf even broader regions.

The end of the superpower rivalry between the US and the USSR has resulted in a number of local and regional military conflicts that were unthinkable at the time of superpower checks and balances during the Cold War era. The demise of global concern over human rights violations has led to numerous ethnic incidents in the area encompassed by the former USSR. In a similar fashion to the fear of nuclear war, regional

[13] Brock, 1992: Security through defending the environment: An illusion? In E. Boulding, Ed., *New Agendas for Peace Research: Conflict and Security Reexamined*, Boulder, Colorado, Lynne Reinner Publishers, pp. 79–102.

[14] R. Moss, Environmental security? The illogic of centralized states responses to environmental threats, *Geopolitical Perspectives on Environmental Security*, Cahier 92-05, Université de Laval, Studies and Research Center on Environmental Policies, GEPRE, Quebec, Canada.

[15] M. Tolba, O. El-Kholy, E. El-Hinnawi, M. Holdgate, D. McMichael and R. Munn, 1992: *The World Environment 1972–1992*, Chapman and Hall, UNEP.

and global concern over the health of the environment should be strengthened to avert the outbreak of environment-related regional conflicts that are already brewing.

A suggestive list of the issues to be tackled, when expanding the notion of security to encompass environmental considerations, are as follows: research to identify the specific impacts of GEC on various countries and regions, separating natural causes of GEC from human-induced ones; environmental factors that contribute to international insecurity; case studies of environment-related disputes and their effects on national and international security and on traditional views of sovereignty; international economic relations and environmental security; the development of regional and global environmental security strategies; designing a system for risk management (for example, in the territories that made up the former Soviet Union).

Key Points

- The environment has become a legitimate national security issue. Thus, a new definition of security must include, in particular, an appreciation of the importance of resources and the environment as key components of national or regional security.
- Four major factors that affect international security in the case of global environmental change (GEC) are: differing perceptions of GEC by government leaders, the uneven impact of GEC on countries, differing contributions of countries to GEC, and a disparity between national or regional capacity to respond to adverse GEC.
- Some ostensibly domestic problems become global policy issues, when other countries take a special interest in them (e.g., tropical deforestation in the Amazon, pollution of Lake Baikal).

WATER RESOURCES

Climate Change and Water Resources

Luis V. da Cunha

Climate plays an essential role in the quantity and quality of regional water resources. According to atmospheric general circulation models, a doubling of CO_2 in the atmosphere could result in an increase in global average temperature of about 3°C bringing about changes in precipitation and evaporation and, thus, also in runoff. The magnitude of such changes in water resources as well as changes in their distribution are expected to have significant social, economic and environmental impacts.

Through the workings of the hydrologic cycle, climatic change will have consequences for vegetation, soils, and more generally the environment. The following figure provides a simple view of the linkage between climate and water resources. Those familiar with the actual workings of the climate system appreciate the complexity of the linkages between climate change, climate variability, and water resources. These conditions will vary considerably from region to region. Thorough investigation and understanding of these linkages requires an interdisciplinary, intersectoral, and, frequently, an international approach.

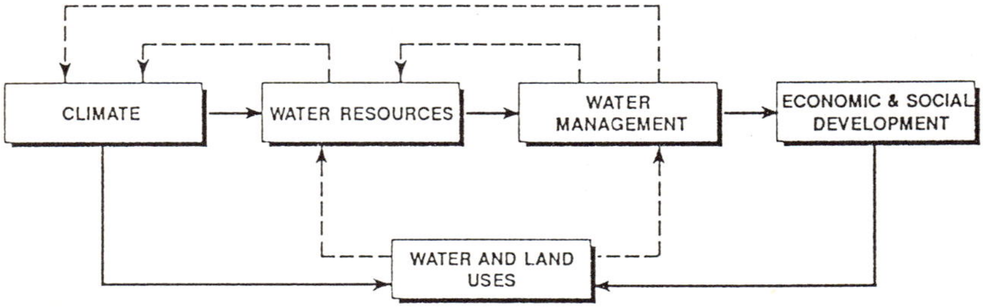

Fig. 1 Continuous lines depict the main relations and dotted lines depict the feedbacks.

Social and economic development can be strongly affected by the quantity as well as the quality of available water resources. In the future, increases in population and standard of living will increase societal demands for water as well as intensify human activities and land-use practices that can also affect the hydrologic regime. For example, changes in agricultural practices, deforestation and urbanization, as well as the construction of engineering works such as dams, canals, or reservoirs interfere to a considerable extent with surface flow. The public and policymakers are more familiar with present-day human-induced alterations in the regional hydrologic regime than they are with hypothetical future climatic changes. However, the consequences of today's climate interactions with human activities are also not very well understood.

Thus, it is important and useful to address the potential magnitude of climatic change impacts on regional water resources. This will enable researchers to identify possible difficulties water managers might face, and to identify ways that current water resources management and planning strategies might be modified to mitigate, if not prevent, the adverse impacts or to take advantage of favorable ones that might be associated with a climate change. When designing dams and other hydraulic works, engineers assume that the climate is stationary, that is, that climate during the next several decades will not differ significantly from that of the last several decades. However, the impacts of a climate change on water resources will not only depend on changes in climatic mean values, but also on changes in climatic variability. The aspects of climate variability that tend to attract the attention of policymakers and the public are changes in the frequency, intensity, and location of extreme events such as droughts and floods.

Engineers do, however, consider uncertainties related to the stochastic nature of hydrologic variables such as runoff. They design hydraulic works to cope with a certain range of climatic fluctuations.

Apart from structural measures to cope with streamflow variations, such as the construction of dams and dikes, there are non-structural measures as well. These include but are not limited to slowing down the rate of global warming through energy conservation, reducing the demand for water through pricing schemes, land-use zoning or abandonment in alluvial valleys subject to floods or coastal sea level rise.

Thus, water resource managers have several non-engineering means available to enhance the capacity of existing hydraulic systems to withstand many of the adverse effects of climatic changes. All these measures notwithstanding, difficulties will be created for water resources planners if the variations associated with the projected climatic changes are significant compared to variations associated with the stationary stochastic process.

Climate change will affect inland water resources (influencing the amount of water available and its geographical, seasonal, and interannual distribution) as well as coastal waters (because of sea level rise and storm surges). Some important consequences of climate change related to inland waters are the following: (a) changes in the spatial and temporal distribution of these resources; (b) changes in soil moisture; (c) changes in the frequency and intensity of extreme events, i.e., floods and droughts; (d) changes in water quality; (e) changes in the sedimentation processes; and (f) changes in water demand.

With respect to sea level rise, increases in sea level can bring about such major modifications in the coastal environment as beach erosion or inundation; submergence of barrier islands, wetlands, mangroves, fish habitats; salt-water intrusion into freshwater aquifers, and so forth.

The interactions between climate, water supplies and management, land use, and economic and social development are quite complex. The most critical effect of climate change is likely to be on the timing and magnitude of runoff and fluctuations in soil moisture at the regional level. As yet, the uncertainty that accompanies GCM-based water resources scenarios makes them difficult to use in the development of realistic management strategies.

The scientific community cannot yet predict with confidence the location or magnitude of regional climatic change. When changing one's analytic focus from the global to the regional level, one should pose the following questions:

- What are the regional climatic scenarios developed for the regions under consideration? Are these scenarios consistent?
- How do these scenarios affect the components of the hydrologic cycle, particularly precipitation and evapotranspiration?
- What are the consequences of these potential changes in precipitation and evapotranspiration for various water resources uses?
- How should the present strategies of water resource management and planning be changed in order to cope with the potential but as yet unknown consequences of future regional climatic changes?

- What are the research needs to resolve the existing problems in the areas of regional changes in climate, hydrology, and water resources management?

Case studies can be an instructive approach to gaining a glimpse of the future. Although no two regional situations are alike, each individual study can help to identify potential problems that might emerge in analogous situations. Each study can help to understand processes related to the allocation of often scarce water resources, as well as in uncovering perceptions about climate variability and change and how those perceptions might affect the water resources decisionmaking process.

A key message to policymakers that emerges from regional climate-related case studies is not to ignore climate. Attention must be paid to the various characteristics of climate, especially its variability and the probability of occurrence of its extremes (including runs of dry and runs of wet years), and not just to climatic means. Climate varies on all time scales, but it appears that decisionmakers tend to focus on climate in the recent past.

The past *is* relevant to climate change research in that it provides a base from which future changes in the mean (and other related climate characteristics) might vary. The problem is not with the direct relevance of the climate of the past, nor with the availability of data (although existing data are not particularly good for many regions of the Earth), but rather how we use that information and how it affects our perceptions about the possibility of future changes in climate. Clearly, there can be costs associated with the lack of understanding of climate information.

Sensitive areas

For many parts of the world, the amount of water available is the factor that limits human activities. This is clearly the case in arid and semiarid areas. A global warming would have major effects on the amount and timing of precipitation in arid and semiarid areas around the globe. Although it is not known (beyond speculation) what the precise effects would be, these areas are hydrologically most sensitive to even small changes in climate. It is not possible to ignore the water resources problems that they face within a season and between seasons. These regions are now tied through the international community to the more affluent (and better-watered) areas of the world.

Finally, it is not difficult to identify the many areas in the world where there are at present, or have been in the past, conflicts over water resources. Such conflicts exist within countries as areas in need of water resources try to deprive other areas of their water through water-diversion projects. They also exist between countries. The United States have had disputes with Canada over the waters of the Columbia River and with Mexico over the Colorado River. Spain and Portugal have had conflicts over various rivers that flow across Portugal to the sea. Central Asian rivers such as the Syrdarya and the Amudarya pass through several newly independent states and irrigate several millions of hectares in the now-multinational Aral Sea Basin. Changes in precipitation in any of the source regions in these examples could adversely affect the amount of water that would be available to downstream areas and lead to possible conflicts of use. Hence, there is a growing concern about the possibility of an increase in "water wars" and perhaps a growing need as well for regional organizations to cope with such potential transboundary resource-related conflict and cooperation.

Key Points

- It is important and useful to address the potential magnitude of climatic change impacts on regional water resources. This will enable researchers to identify possible difficulties water managers might face, and to identify ways that current water resources management and planning strategies might be modified.

- The impacts of a climate change on water resources will not only depend on changes in climatic mean values, but also on changes in climatic variability. The aspects of climate variability that tend to attract the attention of policymakers and the public are changes in the frequency, intensity, and location of extreme events such as droughts and floods.

- The most critical effect of regional climate change is likely to be on the timing and magnitude of runoff and fluctuations in soil moisture at the regional level. As yet, the uncertainty that accompanies GCM-based water resources scenarios makes them difficult to use in the development of realistic management strategies.

- Case studies can be an instructive approach to gaining a glimpse of the future. Although no two regional situations are alike, each individual study can help to identify potential problems that might emerge in analogous situations.

- There is concern about the possibility of an increase in "water wars" and perhaps a growing need as well for regional organizations to cope with such potential transboundary resource-related conflict and cooperation.

Regional Organizations and Climate-Related Changes in the Water Regime

Ödön Starosolszky

Climate change and water resources: Examples from the Danube basin

Very little of the Earth's water is fresh water. Surface water has an average circulation time of 12 days through the hydrological cycle. While total global water resources may be adequate for humankind, their spatial and – especially – temporal distribution are poorly matched to the activities and needs of human populations in different regions around the globe. Diversions and storage of water can often ameliorate some of these temporal and spatial distributional problems. However, the construction of large-scale engineering projects is increasingly being challenged primarily because of its possible adverse environmental impacts.

With regard to importance and order of priorities, societal uses are ranked as follows: drinking water, agriculture, industry and power generation. The water quality requirements are, in general, in the same order. As human populations and their demands for water increase, anthropogenic and natural factors tend to interact more frequently and more obviously result in deteriorating water quality. Water quality issues are particularly important with respect to transboundary rivers. They raise such concerns as who pays for monitoring, clean-up, and compensation. Disagreements over terminology (e.g., shared vs. international vs. transboundary water resources), particularly among upstream and downstream countries, can also be major sources of conflict (and lack of success in common understanding) for most regional negotiations.

Hydrologists have only recently become interested in global warming and its possible impacts on water resources and society. The projected changes in extremes such as maximum and minimum flows will have significant implications for the design of hydraulic structures, particularly because runoff is a nonlinear process. An increase in rainfall intensity would increase erosion, resulting in changes in deposition, which would, in turn, adversely affect water storage capacity and would increase contamination of both water and sediments.

If regional organizations are to respond to the potential consequences of global warming, they will need to increase their effectiveness. The development of new regional organizations may be needed. The experience of regional organizations in the Danube

River basin might serve as a relevant example in designing new regional organizations or restructuring existing ones. In fact, there have been three Danube River commissions. The first was established in 1856, and lasted until the outbreak of the First World War. The second commission existed from 1920 to 1942, and the third was established under the Belgrade Convention in 1947 and continues today. While the second considered general problems of water management throughout the basin, the third has assumed a much narrower scope focusing on navigation, hydrometeorological observations, and regulation involving only riparian countries.

The mandate of today's Commission considers only navigation, and does not take into account environmental factors such as ecology and water management (including water quality). However, two new conventions on these issues are being prepared by the riparian countries. These recognize that minimum flows could be further decreased by greater consumption in riparian countries. Climate change is another factor that could be considered. In past decades, the severity of the impacts of both floods and low flows has increased. Some climatologists have predicted that climate change could result in an increase in the Danube basin in the frequency and intensity of extreme meteorological events and consequently in related extreme hydrologic events. In order for these draft conventions to be implemented, the Belgrade Convention would have to be modified or a completely new organization, involving catchment as well as riparian countries, may be necessary. At present, research is being undertaken by a number of international organizations, often jointly. One focus has been on flood plain vegetation, which has been considerably modified by human activities. Another is on changes in surface/groundwater relationships and in groundwater flow near the river bed, which are more difficult to measure than surface flow.

Proposals for actions

The following activities for regional organizations concerned with water resources, particularly river commissions, may be relevant:

- to predict the impact of a potential climate change on the quantity and quality of regional water resources, especially on transboundary rivers;
- to consider measures to reduce or eliminate the adverse effects of regional changes in the hydrologic regime;

- to coordinate efforts to ensure that water needs are met and to reduce the risks of catastrophic events;
- to harmonize actions to avoid conflicts over water, particularly between upstream and downstream countries sharing transboundary rivers and aquifers;
- to develop and apply water-saving technologies and effective waste-water treatment technologies throughout a region in both upstream and downstream countries;
- to review and, if appropriate, revise multilateral and bilateral agreements with regard to the quantity and quality of transboundary flows;
- to develop international monitoring systems (e.g., data acquisition and transmission) and forecasting and warning systems which may help to trigger protective measures in the event of water-related emergencies (e.g., floods, chemical spills) in both the upstream and downstream countries;
- to undertake pilot projects on selected river basins (e.g., over the Danube and Rhine, within Europe) to demonstrate the potential role of regional organizations in the mitigation of harmful effects that would likely accompany changes in the regional water balance that might result from global warming;
- to formulate and disseminate a statement based on experience gained from previous events about the need for preparedness for critical situations in certain regions and subregions;
- to introduce consideration of the potential implications of climate change into environmental impact assessments of big hydro-projects that would likely affect the water regimes in downstream countries, and into design values and procedures for the existing major water management schemes;
- to improve assistance by donor organizations to mitigate the adverse regional impacts on water resources of climate change (e.g., by constructing dams, diversions, reservoirs, etc., when deemed appropriate and necessary).

Conclusions

The adverse consequences of climate change at the regional and subregional levels, in the absence of proper actions, could cause serious difficulties for a great number of countries. Therefore, actions like those suggested above should be seriously considered. For example, regional organs of the UN system and its specialized agencies should devote more attention to the regional changes that may adversely affect those regions already considered to be extremely vulnerable (e.g., Tigris-Euphrates, Ganges-Brahmaputra, Nile, Niger, La Plata), where changes in water regimes will likely generate serious political or economic conflicts among riparian countries.

Key Points

- The experiences of existing regional organizations such as the Danube River Commission can serve as relevant examples either for designing new regional organizations or restructuring existing ones to cope with the possible consequences of global warming.
- More attention should be devoted to the regional changes that may adversely affect those regions already considered to be extremely vulnerable (e.g., Tigris-Euphrates, Ganges-Brahmaputra, Nile, Niger, La Plata), where changes in water regimes may generate political or economic conflicts among riparian countries.
- Assumptions about climate must be made more specific in discussions about hydrological property rights, or each side will adopt the scenario that it finds most advantageous for negotiation from its point of view.

Climate Change and International Water Problems: Issues Related to the Formation and Transformation of Regional Organizations

James L. Wescoat, Jr.

Problem statement

Research on the implications of climate change for international water management, as an emerging field of study, has given little attention to the role of regional intergovernmental organizations in addressing, resolving, or exacerbating water problems.

In the last few years, a number of major international meetings were held, which one might have expected would consider the role of regional organizations in water resources management. Some of those meetings are as follows:

1) *The Dublin International Conference on Water and the Environment* (January 1992). The Conference participants proposed that management occur at the lowest level possible, with less command and control, and more reliance on market and participatory approaches. The conference recognized the importance of capacity-building at sub-national scales, and that regional organizations are essential for the management of transboundary basins, but it gave emphasis to the former.

2) *The United Nations Conference on Environment and Development (UNCED)* (June 1992). UNCED's Agenda 21 includes at least two relevant chapters, on freshwater resources and on regional organizations. Other chapters, including that on mountains, also mention the role of regional organizations in water resources management and capacity-building. The water resources chapter stresses the importance of both intra-national and international cooperation (including research networks). However, the section on the need for watershed-based organizations remains in brackets (i.e., was not fully endorsed). Individual nations retain responsibility for climate and related water resources impacts. In the chapter on regional organizations, their role is seen as circumscribed, because primary responsibility for environmental management is at the national level. Regional organizations are seen as having a role in the harmonization of approaches, primarily through the use of temporary staff and consultants.

3) *General Agreement on Tariffs and Trade (GATT)* (Uruguay Round, 1992). This includes no references to regional water organizations.

4) *Meeting of the UN International Law Commission* (1991). The draft articles on the Law of Non-Navigable Uses of International Watercourses do not mention regional organizations.

In summary, regional organizations are seen by the international policymaking community as important, but their actions are highly limited by the management of water resources at national and subnational levels.

In preparation for this workshop, Wescoat raised some basic questions about the relationship between regional organizations and international water management. He outlined several comparisons that might be drawn between the water and fisheries sectors in different regions. He then turned to the more difficult analogies that involve regional organizations, climate impacts, and water resources adjustment.

Regional organizations and international water management

During the past four decades, a broad literature has dealt with the notion of "water for peace" and to a lesser extent of "water wars." There have been several important regional experiments with regard to the amicable management of transboundary rivers (e.g., the Mekong and Indus basins), and there have been proposals for new regional organization experiments in the Jordan and Euphrates basins.

These few examples should be viewed within the broader field of "modern" international water agreements and organizations over the past two centuries.[1] A broad perspective raises basic questions about the relationship between regional organizations and international water management, such as the following:

1. *Context & Formation of International Water Organizations*:

   a. To what extent is the formation of major new water organizations contingent upon serious geopolitical conflict?

   b. Are major new organizations more commonly formed to avert conflict, or are they formed in the aftermath of territorial restructuring?

   c. When does the availability of finance capital lead to the formation of major new organizations (e.g., Mekong Committee), and when does it lead to major water development programs with a low level of regional organization (e.g., the Indus basin)?

   d. How do organizations formed for cooperation on substantive water issues in a peaceful situation differ from those indicated above (e.g., in structure, function, power, and performance)?

2. *Transformation of Existing International Water Organizations*:

   a. Under what circumstances do the authority and scope of regional water organizations broaden over time (i.e., from boundary demarcation and navigation to water supply and pollution control)?

---

[1] See FAO, *Systematic Index of International Water Resources Treaties, Declarations, Acts and Cases by Basin*, Legislative Study no. 15, Rome (1978); and Volume II, Legislative Study no. 34, Rome (1984).

b. Under what circumstances are organizations "punished" or abolished for attempting to broaden their scope?

To respond to these questions requires some discussion about the global political, economic, and environmental contexts of regional organization. It also requires some discussion of frameworks for comparing regional organizations in different places and at different times.

Lessons from different regions, sectors, and times

Comparing international water organizations in different regions of the world remains at a rather rudimentary level in terms of theory, method, and practice. As someone who has worked primarily on water issues in North America and South Asia, Wescoat noted that he was particularly interested in comparing the role of organizations in those regions with:

1. International water organizations in Europe (e.g., the Rhine and Danube Commissions).
2. Fisheries organizations in different regions (i.e., what sorts of creative analogies might be drawn from the fisheries sector for water resources management?).
3. International political organizations, for whom the water sector is of negligible or, at most, limited importance.

After comparing recent literature on international water programs with the historical evolution of multilateral water development during the 19th and 20th centuries, Wescoat concluded that perspectives formulated in the 1970s and 1980s may not be especially relevant for the 1990s. What, then, is the "current situation," and are there useful historical analogies to guide regional initiatives?

Implications for climate and related environmental changes

Some of the comparisons suggested above have been, or could be, extended to issues of regional adjustment to climate change. Recent research on "forecasting by analogy" has contributed to the construction of climate change scenarios, assessment of potential impacts, and identification of adjustments that might be overlooked or oversimplified by other methods.[2] These studies have also pointed out the difficulties of constructing

[2] Michael H. Glantz, ed., 1988 ; *Societal Responses to Regional Climate Change: Forecasting by Analogy,"* Westview Press, Boulder.

appropriate analogies. We need to focus in a clear way on some of the following major types of analogy:

1. *Historical*: How does previous experience with organizations in a particular region or sector bear upon the future development of a regional organization in that region or sector, under altered climatic conditions?
2. *Inter-regional:* How does organizational experience in one region bear upon organizational adjustment to global environmental impacts in another region?
3. *Inter-sectoral:* What does experience gained in fisheries and marine resources management in one region suggest for international water management in that region?
4. *Complex analogies:* What does the experience gained in one sector, region, and historical period suggest for another sector, region, and time?

The history of regional organizations for international water resources management extends back in time at least two centuries. A brief examination of the evolution of such organizations could provide some insights, analogies and perhaps lessons for regional organizations in the future. His presentation employed the following outline:

1) Rhine and Danube River Commissions (1804–1868)
   - 1804: Creation of a unified administration for the lower Rhine under Napoleon.
   - 1815: Creation of a central Commission for the Rhine by the Congress of Vienna. This provided for the freedom of commerce and navigation for signatory states, not all of which were riparian: trade and tariff issues were of critical importance.
   - 1856: Treaty of Paris creates a unified administration and commission for the Danube River, after the Crimean War.
   - 1868: The Treaty of Mannheim permitted free access to the Rhine.
2) Colonization of Asia and Africa (1830s–1905). In this period, both territory and trade were critical issues on large river systems.
   - India: exclusive navigation privileges within the Empire.
   - China: bilateral treaties for treaty ports.

- Africa: the 1885 Treaty of Berlin provided for free access for trade on the Congo and the Niger, and established an International Commission for the Congo.

3) European Commissions (1880s–1914):

Agreements and related organizations on Europe's large rivers (e.g., the Danube, the Rhine) began to consider not only territory and commerce, but also pollution, fisheries, transport of hazardous materials, and joint hydraulic works.

4) International nongovernmental organizations (NGOs) (late 19th and early 20th centuries):

   - 1879: The creation of the International Meteorological Organization (now the UN's World Meteorological Organization).

   - 1905: The creation of the International Institute for Agriculture (now the UN Food and Agriculture Organization).

5) The League of Nations and the Pan-American Union (1919–1940):

   - The League of Nations: Programs for water resources management in the League's mandate territories (e.g., Tanganyika) failed. However, flood control on China's Yangtse River was initiated.

   - The Pan-American Union: A multipurpose treaty of water resources (water supply, agriculture, power) was proposed in 1933, but not implemented.

6) Early UN and World Bank programs (1945–1960):

   - Integrated river basin development programs (agriculture and hydro-electricity). Those supported by the World Bank were generally expected to be self-financing, usually through the sale of hydro-electricity.

   - Regional commissions (e.g., ECAFE).

   - Specialized agencies.

   - Expanded Technical Assistance Program (ETAP).

   - Mekong Commission (1957): strong organization, but limited actions due to regional conflicts including the Vietnam War.

   - Indus Commission (1960): weak organization, but well-financed.

Many of these regional programs and initiatives were established in areas of potential geopolitical conflict and were considered to be of geostrategic interest.

Some general lessons can be drawn from this historical experience.

- Major new regional organizations are generally established in the aftermath of a period of political restructuring such as in post-war periods. The post-Cold War period may be just such a period.
- The scope of activities (i.e., functions) of regional organizations tends to change over time. During times of peace, well-designed regional organizations tend to expand.
- Support from individual member states tends to decrease when states feel threatened by regional organizations' activities.
- Conservative regional organizations survive because they abide by the restrictions set out in their constitutions.
- Commerce can be a major impetus to the establishment and survival of regional organizations.

Key Points

- Recent global environmental policy initiatives (including UNCED, the International Conference on Water and Environment, and the International Law Commission) have given surprisingly little attention to the role of regional organizations in international water management.
- Examination of the historical evolution of regional organizations in international water resources management sheds light on their potential roles, constraints, and adjustment to emerging problems such as global environmental change.
- One plausible explanation for the current neglect of regional organizations in international water policy arenas is that new regional organizations tend to be formed during periods of territorial *integration* (e.g., in the aftermath of war or the formation of political and economic alliances). The post-Cold War period may eventually lead to territorial (and thereby water resources) integration in some regions, but not in others experiencing territorial fragmentation.
- Recent research on "forecasting by analogy" has contributed to the construction of climate change scenarios, assessment of potential impacts, and identification of adjustments that might be overlooked or

oversimplified by other methods. We need to focus in a clear way on some of the following major types of analogy: historical, inter-regional, inter-sectoral, and complex analogies.

- In summary, regional organizations might play an important role in international water resource policy, but they will continue to be highly circumscribed by water management at the national and subnational levels.

Regional Organization for Water Utilization in the Middle East

Ilter Turan

Although using water for such purposes as irrigation, municipal, and industrial water supplies has existed for a long time, the magnitude of use was not such that rivers were turned into dry river beds, or deserts into citrus groves. It is mainly during the twentieth century that human beings developed the technologies and accumulated the capital to modify nature in major ways and to bring about substantial changes in their physical and climatic environment.

The rapidity with which such technology-driven transformation has taken place has meant that the short-, middle- and long-term effects of these changes on nature, climate and the environment have been studied neither extensively nor comprehensively. Many problems caused by human-introduced changes are becoming apparent only gradually. Many water systems which have been altered by the transformation have been transnational; that is to say, what has happened in one country has affected what has happened in other countries. Owing to the speed of change, international law sufficient to provide reasonably clear guidelines in the utilization of transboundary waters has not yet evolved. Similarly, there has been a paucity of international organizations of a global or regional nature whose function it is to study, plan, regulate and administer and even adjudicate the utilization of transboundary waters.

It seems particularly difficult to get independent states to cooperate as regards to the use of transboundary waters. First, each riparian considers it an indispensable part of its national sovereignty to use freely and as it sees fit all the waters which flow through its territory. Second and equally significant, countries display an unusual sensitivity to being in full control of their water resources, because being reliant on other countries for water is perceived to be much less acceptable than being dependent on other countries for other commodities such as oil. Third, certain countries tend to benefit from the status quo more than others. Therefore, they are reluctant to take part in schemes which are likely to reduce the advantaged position which they currently enjoy. Fourth, water constitutes only one of the items which affects relations between states. If the countries through which transboundary waters flow are not of friendly disposition to each other and trust is lacking between them, it appears to be very difficult for them to cooperate on water utilization which, for the three previous reasons cited, does not lend itself to being negotiated about in any case.

The difficulties encountered in getting states to cooperate on questions of transboundary water utilization should not allow us to ignore the need for such cooperation. Some observers, in efforts to impress upon both publics and policymakers the importance of securing regional cooperation in the utilization of transboundary waters in the Middle East, have talked about the next wars in the region as breaking out because of water and they have coined expressions such as "playing with water is playing with fire."[1] While it is not prudent to deny the possibility that wars in the region can be also caused by water-related questions, such dramatization may encourage the countries in the region to become even less cooperative than they are today, either for fear that they would be asked to make concessions to avert the danger of war, or alternatively, they might feel that they could be more defiant since going to war is considered a matter of last resort.

In searching for ways that regional cooperation may be enhanced through regional organizations, it may be useful to begin by identifying or delineating the areas in which cooperation is to take place.

It seems that the first area in which cooperation is needed is data collection. In the Middle East, and possibly in other regions of the world where there are transboundary rivers and aquifers, data on water quantity and flows, the size of cultivable areas, the classification of soils according to their potential for effective irrigation, the water needs of population and industry, and other similar variables are not collected using standard criteria and instruments. Furthermore, much of the available data is not brought together in any one place where they are easily retrievable. Much time is lost in negotiations trying to establish the comparability of data provided by different countries. If efforts are made to produce standardized data collection systems and

[1] Dogu Ergil, "Orta Dogu'da Su Savaslari mi?" (Hydraulic Wars in the Middle East?) *Ankara Universitesi Siyasai Bilqiler Fakultesi Derquisi,* vol. XLV, No. 1-4 Ocak-Aralik (January-December) 1990, p. 53-80.

standardized data sets, cooperation might be enhanced between the countries of the region.

The second area in which cooperation is needed is the field of water conservation and the more efficient use of water. To the extent that the unused water supply is decreasing in the entire region and is already exhausted in some parts of it, all societies of the region would have some interest in adopting measures such as turning to water-saving irrigation technologies, using hybrid seeds which require less water, renovating old irrigation systems and urban water systems to minimize seepage and leakage, and other comparable means which would help reduce the amount of water consumed.

Finally, cooperation is needed in planning the collection, the distribution and use of water. It may be better, for example, to allow each riparian to irrigate its prime irrigable land before allowing the exploitation of marginal areas such as those lands with low-fertility soils, the irrigation of which would cost more and its agricultural yields would be less. Similarly, the use of water by one riparian may produce costs for another for which a system of compensation may be devised.

Each area in which regional cooperation is needed may call for a different type of regional, or larger, international organization. In the case of the first area of cooperation (data collection), a large regional organization whose job it is to standardize (i.e., harmonize) data collection systems and procedures, an organization which initiates data collection efforts and offers data storage and dissemination may be suitable or even preferable. It would be preferable because the larger the membership, the greater the chances that the organization will not be perceived as being partial to this or that country or group of countries which may have conflictual relations with others in the region, and that there will be member countries which may help implement and monitor data collection efforts without being suspect of doing, for example, intelligence work on the side for some country which another country in the region sees as an adversary.

In the case of the second area of cooperation (water conservation), we may find a global or an international organization with regional suborganizations the most suitable. Water conservation efforts and the methods of using the existing water supply efficiently are concerns of a universal nature, and the solutions devised are usually globally applicable. Therefore, an international organization that works to diffuse information, knowledge and technology to all members would probably be useful. Furthermore, a universal organization might be able to develop better capabilities for mobilizing financial resources which could be extended to programs in member countries as credits, so that water conservation efforts would not be stymied because of a lack of funds.

In the case of the third area of cooperation (cooperation in the collection, distribution and use of water), smaller basin-wide regional organizations might be preferable. To begin with, political relations between countries in each basin might be different. In some basins, the relations between the riparians may be more friendly (or, at least, less hostile) than others, a condition which would facilitate cooperation and the effecting of mutually acceptable solutions. If several basins in a region are brought under the same organization, the relations between the riparians in a basin in which cooperation proves difficult may affect negatively others between whom cooperative arrangements have a greater chance for success.

Regional organizations which are larger than a basin may elevate problems which can be solved at the basin level to a higher level in which parties that are indirectly or marginally affected by them may try to influence outcomes for ends which have little bearing on the solution of the problem(s) at hand. A corollary is that third parties may get into basin-level matters to pursue exclusively their own interests which may have little to do with the basin as such.

Third, the lack of involvement of non-riparians in basin-wide problems may facilitate solutions, since the external political resources which each party could bring into the problem-solving process would be less, thereby forcing each riparian to settle issues on basin-wide terms. In all likelihood, this would lead to more stable and long-lasting solutions. A solution might not prove as easy to achieve, however, if there is outside involvement, because outsiders are likely to be motivated by reasons of their own which may easily change over time, destabilizing the *modus vivendi* which they might have been instrumental in creating.

Clearly, climatic change is not the only source of change in the distribution of water resources at the regional scale. As noted earlier, most countries in the Middle East regard water as a resource in which they must be self-sufficient. Water availability is a priority of the highest order, and having to rely on other countries is generally viewed as an unacceptable option. A typical approach is to ensure as large a national supply of water as possible, and then decide how to use it. For these reasons, regional organizations for water resources management are lacking in this particular region, and there is great resistance to change.

Excluding the Nile, the Middle East includes four major river systems. While some rivers are barely flowing in some seasons, all sources of water are important in this region of water scarcity. Very briefly, some of the important issues in these basins are as follows:

- *The Jordan River*: completely allocated; groundwater is used extensively. Yet, water-based agriculture continues to be promoted. The Golan Heights probably carry as much importance as a water source as they do for military strategic reasons.
- *The Litani River*: completely allocated. Israeli security zone reaches this river, raising suspicions as to whether there might not be a hydrologic dimension to this military intervention.
- *The Orontes River*: In the past, Syria has had claims on the Turkish part of the basin. Syria tries to control the waters of the river as a counter-measure to Turkey's possible trapping of the waters of the Euphrates.
- *The Tigris-Euphrates*: these two rivers carry enough water for the current needs of the riparians. Syria and Iraq have become concerned that Turkey, which had used the waters of these rivers mainly for the generation of power in the past, has recently completed the construction of the Atatürk Dam on the Euphrates which will irrigate large areas. Syria and Iraq did not want to be involved in a joint project aiming at the construction of a dam in Turkey, and subsequently blocked World Bank loans to Turkey. Turkey managed to build the dam with its own means. The fear that Turkish plans to develop a string of new irrigation dams on the Tigris and the Euphrates will lead to its consumption of too large a quantity of water, not leaving enough for the downstream users, has served to exacerbate tensions deriving from other causes between the three countries. The conflict carries the potential of hardening into a Turco-Arab polarization in the region.

Some generalizations emerge with regard to water resources management in the Middle East region:

- Population and irrigation-driven agricultural development are increasing in the region. The most efficent use of existing supplies is not given sufficient attention in the region with the possible exception of Israel.
- Although in the foreseeable future, water quality may become just as important an issue as quantity on some rivers such as the Euphrates, most assessments of regional water resource focus exclusively on quantity.
- Most countries do not favor international cooperation either because they benefit from the status quo or they hope to change it by unilateral acts.

- The predominant trend of individual national action leads to collective losses. As noted earlier, there appear to be roles for regional organizations in the region in the realm of standardization of measurement (e.g., water discharge), classification (e.g., soils), and regulation (e.g., pollution control). The collection of standardized data might constitute a beginning point on which further cooperation might be built.
- Depending on the nature of the problem to be dealt with, different types of organizations might be more suitable, as explained in the earlier discussion of cooperation schemes in different areas.

Regional organizations should not be viewed as a panacea which would bring solutions to water problems and the ensuing problems of environmental and climatic change. The idea which I have tried to emphasize is that different needs may call for different types of regional organizations. There is no standard organizational type which is suitable for dealing with all aspects of water, environment and climatic change issues at the regional level. The choice of an inappropriate type of organization may impede rather than enhance the implementation of solutions.

Key Points

- Climatic change is not the only source of change in the distribution of water resources at the regional scale. Most countries in the Middle East regard water as a resource in which they must be self-sufficient. Water availability is a priority of the highest order, and having to rely on other countries is generally viewed as an unacceptable option.
- It seems particularly difficult to get independent states to cooperate on the use of transboundary waters despite the need for such cooperation, because countries display an unusual sensitivity to being in full control of their water resources.
- It may be useful to first identify those areas in which regional cooperation through regional organizations is to take place, such as data collection, water conservation, and the more efficient use of water.
- A regional organization should not be viewed as a panacea which would bring solutions to regional water problems and the problems of environmental and, possibly, climatic change.

Transboundary Water Resources on the Iberian Peninsula

Rui Roda

Introduction

More than 40% of Portugal's total surface water supply originates in Spain, mainly through the Douro, Tagus and Guadiana rivers. Its availability depends on the exploitation of the existing Spanish reservoirs. The conventions between Portugal and Spain cover only the hydroelectric use of international rivers and hydraulic works. As a downstream user, Portugal has been concerned about the increase in the demands of Spanish upstream water users and the related threat to water quality.

Geographic and climatic situation in Portugal

The Portuguese mainland is located in the western part of the Iberian Peninsula with an area of about 89,000 km^2. Its climate is defined by its geographic location through the influence of the Atlantic Ocean (mainly in the northwest areas in winter) and of the Mediterranean Sea in the south and east parts of the country in summer. Portugal's annual average precipitation is about 900 mm/year, ranging from about 3,000 mm in the north to 340 mm in the south (the European annual average precipitation is 734 mm). Precipitation on the Portuguese mainland takes place during autumn–winter–spring seasons and is marked by a high level of interannual variability.

International river basins in Iberia

The following figure shows the major river basins on the Iberian Peninsula. It is important to stress that (1) more than 60% of the Portuguese territory is encompassed by Portuguese–Spanish watersheds; (2) the Portuguese area of the Portuguese–Spanish watersheds corresponds to 22% of the total; and (3) in Portugal, the maximum available surface water resources reach about 21,000 million cubic meters, of which approximately 40% come from Spain.

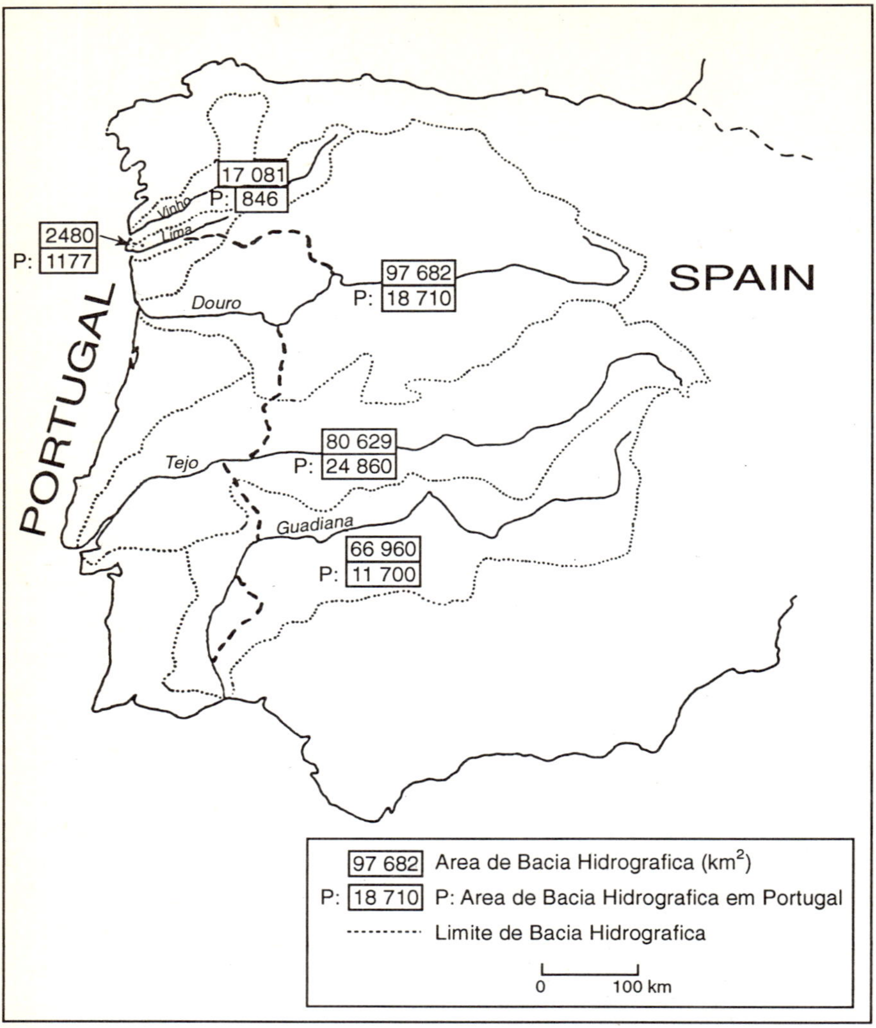
SPAIN
PORTUGAL
17 081
P: 846
Vinho
Lima
2480
P: 1177
97 682
P: 18 710
Douro
80 629
P: 24 860
Tejo
Guadiana
66 960
P: 11 700
97 682 Area de Bacia Hidrografica (km²)
P: 18 710 P: Area de Bacia Hidrografica em Portugal
Limite de Bacia Hidrografica
0
100 km

Climate change

Scenarios about possible impacts of climate change in southern Europe suggest that precipitation will decrease in all seasons except autumn, when there would be an increase. This potential decrease in Iberian precipitation in the spring and summer seasons would generate a major crisis in Portugal, considering existing present-day chronic water shortages. For its part, an increase in precipitation in the autumn would increase the risk of floods and would intensify soil erosion.

Need for cooperation Between Portugal and Spain

The hydrologic year which just ended (Fall 1992) was characterized by extremely dry winter and spring seasons. The reduction in the quantity of available water also accelerated the degradation of its quality. This recent situation provides insights about socioeconomic difficulties which could occur in the future. It also reinforces the urgent need for joint action between Portugal and Spain, for the future well-being of all the inhabitants of the Iberian Peninsula.

To sustain present water consumption levels, it will be necessary to do the following: (1) to increase the water storage capacity for improved, equitable annual and interannual regulation; (2) to improve flood protection and flow regulation; (3) to improve water quality.

Investment in regional water resources development, however, will only be effective if accomplished in an integrated fashion by the joint efforts of Portugal and Spain. Advantage must be taken of the Peninsula's physiographic conditions and of the joint capacity to mobilize financial resources, namely the EC's structural funds.

In addition, the resolution of future climate-change-related problems cannot be based solely on substructural solutions designed only to increase water availability. Spain and Portugal must also act to increase the efficient and equitable use of water resources. Water regulation and management must be improved through a more equitable sharing between Spain and Portugal. Such joint cooperation could develop within the present cooperative framework for hydroelectricity and hydraulic works.

Key Points

- The conventions between Portugal and Spain cover only the hydroelectric use of international rivers and hydraulic works. As a downstream user, Portugal has been concerned about the increase in the demands of Spanish upstream water users and the related threat to water quality.

- The hydrologic year which just ended (Fall 1992) provides insights about socioeconomic difficulties which could occur in the future. It also reinforces the urgent need for joint action between Portugal and Spain.

- The resolution of future climate-change-related problems cannot be based solely on solutions designed only to increase water availability. Spain and Portugal must also act jointly to increase the efficient and equitable use of Iberian water resources.

MARINE RESOURCES

Some Aspects of Regional Cooperation in Marine Sciences

Gunnar Kullenberg

Background and UNCED

Regional cooperation in marine research and systematic observations is of long standing. One example is the International Council for the Exploration of the Sea (ICES) initiated at the end of the 1800s and the intergovernmental agreement formally concluded in 1902, only about 25 years after the meteorologists started their global cooperation in the International Meteorological Organization (now World Meteorological Organization). Several decades later, a milestone in the development of ocean affairs was laid down in the Santiago Declaration of 1952, announcing the then-controversial 200-nautical-mile extended economic zone limit for the Comision Permanente del Pacifico Sur (CPPS), an organization created at the same time. This has gradually led to an extensive cooperation of marine research and observations in that region, founded on a clearly identified need related to renewable marine resources of overwhelming richness.

Cooperation in research in the North Atlantic led to regional cooperation in fisheries management through a series of regional fisheries management commissions. This structure included active participation and advice from the scientific community. The CPPS region worked toward the same end, but started from the perspective of management and jurisdiction rather than the need for cooperative research.

The preparations for UNCED 92 (Earth Summit) emphasized the regional scale in the context of Agenda 21 for the Oceans. It has been considered viable to enhance regional cooperation and coordination for the development, protection and sustainable use of the marine environment and its resources. At the same time the legal and management framework provided by the UN Conference on the Law of the Sea (UNCLOS) has been emphasized. The mechanisms provided through the Intergovernmental Oceanographic Commission's (IOC) regional subsidiary bodies can, in close cooperation with other parts of the UN system and in particular UNEP, provide the basis for the implementation of studies and systematic observations required for proper assessment of available resources, including their distribution, for marine environmental audits, for impact assessments of climatic and other changes, for formulation of control measures, and for help toward their enforcement.

A proper information base and fact-finding process must be part of any plan to achieve sustainable development. With proper coordination and dialogue this can be achieved through existing mechanisms, provided that the coordination and dialogue at regional levels are matched by similar efforts at national levels. It should be pointed out that coordination of mechanisms and cooperation between the existing bodies dealing with the marine environment has been rather good. It includes frequent interagency consultations, agreements with the Joint Group of Experts on Scientific Aspects of Marine Pollution (GESAMP) and the Intersecretariat Committee on Scientific Programmes Related to Oceanography (ICSPRO), and interactions within and between regional bodies.

The IOC started fostering regional cooperation from its inception by supporting and organizing at the intergovernmental level the Indian Ocean Expedition in 1960–65. This was a regional endeavor carried out with a global perspective and with global support. It has, to some extent, become a guideline for regional cooperation.

The regional subsidiary bodies of the IOC have been developed since the middle 1970s, and particularly during the 1980s. The basic ideas are that regional cooperation, including that of marine science, is a necessity, (1) because of the dynamics of the ocean and the marine environment (as well as its resources); (2) because a regional perspective helps define priorities and priority needs; (3) because of the pooling of efforts; (4) because it fosters regional and national development and supports a multilateral economy; and (5) because of the need to address many marine problems on a regional rather than a national or global basis in view of the constantly changing properties of the ocean environment and the ocean's interactions with land and atmosphere. Regional cooperation in marine matters is somewhat easier to achieve than regional cooperation on land because of fewer territorial concerns at sea than on land. Hence, the "regional window" is a natural one as far as marine sciences are concerned. The debates during the formulation of the UN Law of the Sea Convention, and the Convention itself, helped to develop these ideas considerably, mainly through regimes inherent or specific to the Convention.

Perspectives

The lessons learned for the future include: the "identification" and "identity" complex (in the sense that the population and government of a country feel more easily identified with their close environment and related immediate problems, and hence with their "own" or at least, neighboring marine environment) and partnership; the critical mass complex; the need for a long-term (decadal) policy and plan, which include common interests and common priorities; the need for governmental commitments and involvement; the associated need for regional (not external) governance of the region; the stimulation to be obtained from regional cooperation as a driving factor; the creation of IOC regional subcommissions on the basis of the start made through regional committees; the cooperation and "supplementarity" of IOC and UNEP regional actions and the gradually increasing interaction of their regional bodies.

Future success lies in the ability to convince support systems and individual government institutions of the benefits of multilateral cooperation which can, in fact, supplement bilateral aid or donor activities. There is also a need for an increase in coordination and cooperation between donor institutions. Without a change in attitude and the development of global solidarity, the aid alone will not suffice.

An important aspect of the future directions in marine science must encompass the integration of open ocean research programs with those traditionally focused on continental shelf seas and in coastal zones. Consideration of open ocean processes in isolation (or separately) from the processes occurring in coastal zones will most likely be counterproductive in the face of global climatic changes and sea level rise. Such integration needs to be taken into account in regional cooperative programs.

Countries from the North and the South, against the background of complicated global and regional marine issues, have gradually realized the need for cooperation and the mutual benefits that can be derived from it. The pooling of resources, the sharing of information, and education and training are all essential elements in the creation of a solid foundation for the global approach that is necessary to narrow major outstanding scientific uncertainties. Multilateral cooperative approaches provide a better basis for making informed decisions about the future and for avoiding additional potentially immense problems for future generations, problems that might result from present-day actions.

Major phases in the development of regional cooperation

As a result of the establishment in 1960 of the IOC, a mechanism was created to foster intergovernmental cooperation in marine science. The IOC, from the start, emphasized regional activities, through concerted action, with the participation of the states, both from inside and outside of the region of concern. Some examples are as follows: the International Indian Ocean Expedition, 1960–65; the Cooperative Study of Kuroshio and Adjacent Regions, 1962–77; the Cooperative Investigations in the western Africa region; and in the Caribbean.[1]

These activities gradually led to the realization that regional cooperation gave great benefit to the participating nations and to the establishment of regional coordinating committees. The negotiations during the UNCLOS stimulated an upgrading of these bodies to Regional Committees as IOC regional subsidiary bodies. The existing IOC regional subsidiary bodies are as follows:

IOC Sub-Commission for the Caribbean and Adjacent Regions (IOCARIBE)
IOC Regional Committee for the Central Eastern Atlantic (IOCEA)
IOC Regional Committee for the Cooperative Investigations in the North and Central Western Indian Ocean (IOCINCWIO)
IOC Regional Committee for the Central Indian Ocean (IOCINDIO)
IOC Regional Committee for the Southern Ocean (SOC)
IOC Sub-Commission for the Western Pacific (WESTPAC)

It is noted that two of those regional committees have been upgraded: the IOC Regional Sub-Commissions for WESTPAC and for the Caribbean and Adjacent Regions (IOCARIBE). These sub-commissions have broader functions and roles than the regional committees, and their elevation suggests a certain regional maturity with regard to marine research observations, and cooperation in these regions. The IOCARIBE Sub-Commission was created in 1984 and the WESTPAC Sub-Commission in 1989–90. These are still parts of the IOC and are not independent bodies. The idea is that the regional programs and their implementation shall benefit from the global program developments in the IOC framework, but that they, at the same time, address regional priorities. The establishment of the sub-commissions was done only after careful, in-depth evaluation of the developments of the region's marine programs.[2]

---

[1] See, for example, Roll, 1979, IOC Technical Series No. 20; Haq, 1989: An evaluation of marine science cooperation in the Western Pacific Region. Document IOC/INF-773.

[2] See e.g., Haq, 1989, op. cit.

The possibility of merging the IOCINDIO and IOCINWIO Regional Committees into one IOC Sub-Commission for the Indian Ocean and Adjacent Regions has been discussed. While this may be premature at present in view of the large differences in the state of development of marine science in the countries surrounding the Indian Ocean, there is no doubt that the creation of an IOC Sub-Commission for the Indian Ocean is a goal for the 1990s.

The increasing international interest in the Southern Ocean in relation to its role in the global climate system, the potential further exploitation of marine resources in the Southern Ocean and its possible contamination by long-range transfer of pollutants as well as a result of activities in the region and in Antarctica, have led to a realization of the lack of data from this region. Efforts are, therefore, under way in the IOC and its Regional Committee for the Southern Ocean to increase ocean observations there; the World Ocean Circulation Experiment (WOCE) has a large hydrographic program in the area. There is a need for increased cooperation in marine research in the Southern Ocean, in view of its role, possibly vital, in the climate system. Regional cooperation is also increasing in the Arctic and northern North Atlantic in the context of the climate system research, and in view of the role of these areas in the functioning of the global climate system. Considerations are under way for the creation of a regional center for logistics and operational support in Iceland. The IOC has expressed interest in cooperating with such endeavor.

One basic IOC task is to facilitate international cooperation in marine science, among member states and among organizations. Hence, the IOC subsidiary bodies are endeavoring to cooperate with other regional bodies inside and outside the UN system. In this way, cooperation and dialogue are established between the marine science community and the management communities who are supposed to use the analyses of marine research and of systematic ocean observations. While such a dialogue is a must, it is nevertheless difficult to establish. For example, while there is some interaction with the regional FAO Fisheries Council and Commissions, it is not very much. The efforts and potential of the IOC regional mechanisms are hardly referred to in the recent study by the World Bank, FAO and others on the development of fisheries and related ocean research.

In contrast, very close cooperation has been established by the IOC with the UNEP Regional Seas Programme in most regions. The UNEP Programme addresses questions related to the management and control of marine pollution mainly from land-based activities, focusing on a regional scale. This program is, thus, interested in using scientific results and adequate, reliable observations. In many cases IOC and UNEP regional definitions overlap, and cooperation and coordination between the programs and activities would, therefore, seem natural. It has, however, taken a long time and strong efforts to achieve this goal. Presently the cooperation is strong in several regions and includes joint programs in the Caribbean, Southeast Pacific, East and West Africa. Such cooperation has not yet been developed in the South or Western Pacific, or in the East Asian or South Asian Seas. This lack of cooperation in these regions can be explained by the different definitions of these regions and to the level of development of the programs, and to the fact that IOC has not really been present in the South Pacific.

The IOC has also promoted a policy of cooperation and coordination with regional bodies, including the International Council for the Exploration of the Sea, the Helsinki Commission, the International Council for the Scientific Exploration of the Mediterranean, the Comision Permanente del Pacifico Sur (CPPS), and the Indian Ocean Marine Affairs Cooperation.

In many cases IOC has entered formal agreements or "memoranda of understanding" with partners, well reflecting the complementarity of their activities, as well as the need for dialogue and cross-sectoral interactions between different sectors of society.

Some experiences

In implementing the priority programs in marine research and services adopted by the IOC regional subsidiary bodies, there have been many successes as well as several difficulties. For example, the regional marine pollution program of IOCARIBE, CARIPOL, developed very successfully through a careful step-wise procedure, which included identifying the problem, focusing on something realistic and do-able, developing the appropriate methods and manuals, instituting the required training, formulating the regional program (including an agree-upon data submission procedure, data analysis and evaluation), bringing together all participants at suitable intervals so as to ensure an active feeling of a network and exchange of experiences, together with the stimulation of seeing and using data obtained from partners in the network. The only gap was the application of the information to formulation and implementation of marine pollution control and abatement measures. This was lacking since there was no adequate dialogue with, for example, the UNEP Caribbean Action Plan, or the International Maritime Organization. This situation has now been remedied, but in the process the close feeling of a network among the CARIPOL participants has been lost. There are several other success stories – but of course there are also failures.

What, then, are the major problems which have been encountered? These include: lack of communication; lack of governmental commitment or priority setting, which made it difficult to obtain, for example, UNDP funding; lack of commitment on behalf of the national institutions, which made it difficult to have active participation of scientists and technicians, since there is no apparent benefit in it for them; lack of funds and partners in the developed regions; and lack of equipment and trained personnel. These problems are interlinked. A basic aim has been, and is, to train personnel – but if the institutions are not committed, then personnel are not available for training. The lack of governmental support is, of course, related to the fact that marine matters usually are not in primary focus – humans are, after all, terrestrial animals. The remedies include an increased governmental and public awareness of the role of the ocean for development and, in that context, demonstration of the importance of marine research and systematic observations through examples of rational use of the whole environment. One such example is provided in the use of the results of the Tropical Ocean Global Atmosphere (TOGA) program. These programs cannot be operated from far away. There is a definite need to have a committed and very competent staff present in the region and actively visiting the participating institutions. One way to achieve this is through the development of regional centers. Functioning national liaison mechanisms need to be established as well, which are linked to all relevant ministries and government departments. These bodies must be active on the national scene and fulfill a coordinating role at the national level.

The mechanism of regional centers for marine science and technology, as foreseen in UNCLOS, should be considered. Such centers may provide the obvious coordination and dialogue mechanism, provided that the structure permits them to work closely with all relevant partners within the UN system, as well as with others outside of it. Isolated centers will not serve the coordination function. The implementation of the IOC/UNESCO Comprehensive Plan for a Major Assistance Programme to Enhance Marine Science Capabilities of Developing Countries could conceivably be tied to the development of such centers, which would inevitably contribute to and establish partnerships among participating member states, both from the region and outside. Such centers may provide a basis for an optimal use of available resources including a scientific community which is often below the critical mass of the national levels, but on a regional integrative basis becomes above critical mass.

Finally, improved coordination among donors and various supporting agencies is necessary. Bilateral and multilateral aid mechanisms must be seen as complementary and not conflicting. Use should be made of a regional mechanism to generate support from several donors, on a bilateral or multilateral basis, towards the same goal or program over a whole region. This is what the IOC is attempting to achieve through its regional subsidiary bodies, by establishing a dialogue with national donor agencies for the purpose of obtaining support for agreed-upon regional programs.

Key Points

- The basic ideas of the IOC are that regional cooperation is a must, because of the dynamics of the ocean and the marine environment (as well as its resources); because a regional perspective helps define priorities and priority needs; because of the pooling of efforts; because it fosters regional and national development and supports a multilateral economy; because of the need to address many marine problems on a regional rather than a national or global basis in view of the dynamical properties of the ocean environment and its interactions with land and atmosphere. Hence, the "regional window" is natural as far as marine sciences are concerned.
- Use should be made of a regional mechanism to generate support from several donors, on a bilateral or multilateral basis, toward the same regional objective. The bilateral and multilateral aid mechanisms must be seen as complementary and not conflicting.

The Potential Role of Regional Organizations Related to the Marine Environment in the Context of Global Climate Change

Edward Miles

My hopes and concerns about the potential role of regional organizations in the context of global climate change are heavily shaped by the experiences to date in responding to other global environmental problems and in seeking to facilitate international collaboration in scientific research, in exploiting while protecting the global commons, and in R&D collaboration in advanced technology. All of these activities have important characteristics in common. Most significantly, they are all mixed motive "games" (i.e., combining conflict and cooperation) and, therefore, fall within the class of collective action or public goods problems where a well-developed theory already exists.

Let us not forget that, for states, international collaboration is never an end in itself. Collaboration is always purposive, the aim being to solve some problems or realize some gains which states could likely not have achieved, left to their own devices. Relative to any set of given conditions, therefore, we must ask as analysts at least two questions: a) will cooperation/collaboration (I use these terms synonymously to imply the search for joint gain) emerge among the states in question? and b) If it does, will it be enough to solve the problem or to realize the gains sought by the "players"?

The type of issue (or problem) is also an important variable. While global climate change is a new issue on the international agenda, it is not a new type of issue (or problem). It falls into the sub-class of global environmental problems of long time scale where uncertainty is high, where scientific consensus about causes and consequences evolves slowly, where trade-offs are potentially painful in a socioeconomic sense, and where political inertia ("do nothing, or at least do very little") is great.

What then can we expect of regional organizations in the context of the global climate change problem? There appear to be important, though subtle, connections between (a) the degree of technical and/or political difficulty of the problem to be solved, (b) the degree of political salience of the problem or how important it is to states to seek a solution, (c) the nature of the regime in which the problem arises, and (d) the relative capabilities of the states dealing with the problem. Even if one is facing really malign problems, as is the case with global climate change, where both technical and political difficulties are great and the regime is somewhat permissive, a fair amount of collaboration can still emerge over time.

For this to occur, if one looks at, for example, the North Sea pollution issue and the long-range air transport of pollution issue in Europe, it appears that capabilities among the players need to be relatively symmetrical and at a high level; the issue must be politically salient to domestic constituencies at home to the extent that governments must pay attention; and the regional organization then emerges as the mechanism wherein learning can occur. For learning to occur, what Peter Haas calls "epistemic communities" appear to be crucial.[1] But, let us not push our hopes too far. There is nothing either automatic or irreversible about this process and, even if collaboration

[1] P. Haas, 1990: *Saving the Mediterranean.* New York: Columbia University Press.

increases, it does not mean that there will be enough collaboration to solve the problem. Moreover, time itself is a critical variable, since it can produce significant changes in the perception or definition of the problems to be solved. One is, therefore, dealing with a shifting target.

Contrast these lessons with what happens when the political salience of the issue is mixed or low and capabilities among the players are either uniformly low or sharply skewed. I will use here as an example the attempt by Dr. Mario Ruivo as Secretary of the Intergovernmental Oceanographic Commission (IOC) of UNESCO to pursue a deliberate strategy of organizational growth based upon (1) the potential role of IOC as an information broker in the context of the new ocean regime, and (2) the development of IOC's regional programs as the "arms and legs" of the IOC central office in Paris. Theoretically, both approaches made sense, although on reflection the value of IOC as an information broker had politically important limits. I confine myself to the regional "arms and legs" strategy, which is of definite relevance to our concerns. It ran aground in the regions for a variety of reasons. In the Caribbean, for instance, IOCARIBE was hampered by the lack of funding to implement projects recommended by the experts; by the absence of funding to support travel for coastal state scientists in the region to attend planning meetings; by the fact that since these scientists were in short supply in the first place, governments were not too eager to assign them to regional activities since they would not then be available to work on other things; and by the fact that the two dominant coastal states in the region (Mexico and Venezuela) did not seem to rank IOCARIBE activities high on their own scale of priorities.

As a second example, in the Central and Western Indian Ocean, the mechanism (CINCWIO) was at first a very effective one, given a change in the world ocean regime which made access to Exclusive Economic Zones (EEZs) for the purpose of conducting marine scientific research as a commodity to be traded. The region was of great scientific interest to advanced maritime states whose scientists wanted to study the impacts of the reversing monsoonal effect on oceanic upwelling. Accordingly, the research that went on under the CINCWIO umbrella allowed advanced maritime states access to achieve their objectives, while they provided coastal states assistance in focusing on regional problems of primary interest to them. But once the advanced maritime states achieved their objectives, the funding to continue work in the region disappeared.

Left to their own devices, the coastal states of the region were divided between those who lacked all capability to engage in marine scientific research and those states in which some capabilities existed, such as Kenya, Tanzania, and the Seychelles. But, even among the latter, one found that political rivalries (between, for example, Kenya and Tanzania and between the Anglophones and the Francophones) greatly constrained the level of collaboration achieved.

In the Western Pacific, Project WESTPAC is perhaps the most viable of these three examples, but this is a case in which differing levels of capability presented difficult problems of managing diversity. For instance, the political salience of the region to the advanced maritime states involved was high, *but for different reasons*. For the former Soviet Union, importance was defined in terms of expanding access for conducting marine scientific research to support resource and strategic objectives of the then-Soviet Government. For the United States, Japan, and Australia the purpose, not surprisingly, was to deny access to the USSR.

For Japan, the region represented an area of historic involvement where foreign policy and resource (fish and oil) objectives loomed large. For France, involvement in WESTPAC was a way of buying political legitimacy in the face of intense regional hostility on issues involving (a) French nuclear testing and (b) independence for New Caledonia. For China, WESTPAC represented a potentially important mechanism for receiving assistance on marine scientific research, but only if China were able to exclude the USSR from automatic access to the Chinese EEZ for purposes of conducting research.

When, in the early 1980s, one probed representatives of all member states to WESTPAC on the *scientific priority* of the region, the results were that it was considered very high by all developing country representatives indigenous to the region and for the USSR. For all other advanced maritime participants, the scientific priority of the region was low. Accordingly, there were real limits on what they were willing to do and it was very difficult to bridge this gap. This was so even when, objectively, WESTPAC was of major potential importance as the sole mechanism in the region facilitating discussion of common scientific problems involving coastal states where, normally, diplomatic problems and conflicts of great sensitivity would have precluded such cooperation. But then, pollution in the Pohai Gulf was not a problem of major significance to the United States, Australia, France, or even to Japan.

I think these experiences are instructive, if one looks at the magnitude of the task in coming to terms with the global climate change problems at the regional level.[2] Here, the mismatch in priorities (salience) and capabilities can be deadly. If the scientific interests of advanced players is even medium to low more often than not, and if their political interests are similar, one can still see regional arrangements being created – but their performance will remain at a low level. Developing country interests and priorities may be high but the lack of resources and infrastructure on their side can never overcome the major hurdle described above.

Let me then stick my neck out and suggest that *regional cooperation works to resolve environmental problems only when the scientific questions are of high priority for advanced players and when no other means of attacking them exists.* Scientific questions become high priority under one or a combination of three conditions:

a) when there are significant material gains to be had;

b) when there are politically significant "bads" to be avoided; or

c) when major domestic constituencies compel governments to face up to the issues, even though the leadership and certain interest groups may not wish to do so.

Leadership at the regional level is important but there are real limits to policy innovation if the problem to be solved is malign, if capabilities are not distributed symmetrically, and if the distributional consequences of the available policy options are all severely skewed. But, even when regional cooperation "works," one should expect an inevitable rivalry between the regional level and the bilateral level, because the costs of interdependence of the latter are always less.

Type of technology can be an important tool in the hands of imaginative leadership if it can be used to extend or constrain the size of the pie to be shared. Similarly, issues/problems can change their nature over time if the divisibility of benefits/costs varies significantly. Above all, one needs to try to design and construct management systems which can learn from trial and error.

---

[2] See, for instance: Gable, F.J., D.G. Aubrey, & J.H. Gentile, 1991: "Global Environmental Change Issues in the Western Indian Ocean Region," *Geoforum*, Vol. 22, No. 4, pp. 401–419.

A focus on fisheries management regimes

A key question in the management of fisheries is who has control over fish stocks (through property rights), control over fishing activities, and a management system with well-defined authority. This requires avoidance of divided authority, a reliance on voluntary participation and management systems, and constraints on surveillance and the enforcement of regulation. Two different regimes for the use and management of marine fisheries can be identified, with a transition period in the 1960s and 1970s.

The old regime can be characterized by the concept of "freedom of the seas," for the exploitation of most marine fisheries, with flag-state jurisdiction and weak surveillance and enforcement. Unanimity was required for management decisions and "tragedy of the commons" situations occurred frequently. This open-access situation was characterized by limited coastal-state jurisdiction, highly divisible benefits, minimal collaboration, and fragmented authority with regard to resource exploitation. There was an absence of transnational management processes, because the fisheries were always viewed as a part of the question of coastal-state jurisdiction; and so governments did not allow non-governmental actors independent access to regional commissions. There were few, if any, incentives to cooperate in this "zero-sum" game. One major exception concerned the capture of marine mammals, where changes in management regimes resulted from the actions of environmental groups, based more on "religious" (i.e., emotional) grounds than on scientific findings. In this case, non-governmental organizations significantly constrained and shaped the actions of a global intergovernmental mechanism: the International Whaling Commission (IWC).

The first international commission for the management of marine resources, the North Pacific Fur Seal Commission (1911), was followed by many others. The purpose of such commissions is the cooperative solution of problems that cannot be solved individually, that is, collective action. However, few commissions have been totally successful, although they have been highly useful, through low- to medium-level collaboration.

Under the old regime, there were management and non-management commissions. The activities of management commissions could be defined as follows: (1) collecting, disseminating, and analyzing information; (2) deciding quotas and allocating catches; (3) undertaking joint research and information exchange. Non-management commissions were usually organized by the UN FAO. They also collected, analyzed, and disseminated information, but usually under more difficult conditions, and occasionally recommended quotas and catch distributions. Their greatest successes were in mobilizing, aggregating, and articulating the interests of developing countries; designing and implementing large-scale technical assistance programs; providing advice on fisheries to the UN; protecting and enhancing their institutional domain.

The new regime has been marked by changes in state practice, with the extension of coastal state jurisdiction from 3 miles to 12, and then to 200 miles in the mid-1970s. This has led to various conflicts, which have been partially resolved by the Third UN Law of the Sea conference, which provided rules for exclusive economic zones (EEZs). Three articles (61, 62, and 297[3]) give to coastal states the unreviewable right to define catch limits, and their own capacity to take whatever yields are available. Thus, property rights within 200 miles of the coast were vested in the coastal state. This has resolved the problem of fragmentation of authority for foreign fleets within EEZs, but not for the domestic ones. Problems of domestic capacity to manage domestic marine resources are now dominant. International problems remain, however, where stocks are still shared between coastal states or where stocks straddle the EEZs and the high seas.

With respect to assessing the impacts of the new regime, there are four critical factors to consider: the nature of the external environment; the nature of the task and supporting technologies and capabilities; the life-history of the species in relation to the capture technology used; and the divisibility of payoffs.

The EEZ regime limits the role of regional fisheries commissions, so that most have lost their former management responsibilities to the coastal states. The commissions can only be effective when their activities are harmonized with the objectives and policies of specific states. Thus, it appears that innovation may only be possible in the cases of shared and straddling stocks. While there are possibilities for increased surveillance and enforcement, these require more and better information and, therefore, additional financial resources. Overall, there are now three major hurdles to overcome: (1) the effects of incompatible development and utilization; (2) the low levels of scientific and management expertise in most developing countries; and (3) the inequities in paying for access.

With respect to climate change, the following four lessons can be learned from past experience with fisheries institutions:

1) The nature of the decision system is critical. At the global level, there is no alternative to the "law of the least ambitious program." If all countries have

one vote, regulations will be watered down in order for them to be accepted by all. With climate change, it may be better to proceed sequentially, starting with industrialized countries, which should decrease their greenhouse gas emissions first and fastest, leaving additional time for research and for permissible growth in developing countries.

2) The divisibility of payoffs is critical. How can regulations be used to ensure equity? How should "pain" be apportioned?

3) Regulation requires monitoring and enforcement, so that the costs of "cheating" are greater than the gains to be derived from cheating.

4) The role of science is crucial. The most credible scientific voice would be a global one (i.e., the IPCC). Science must not be linked so closely to regulation that governments meddle with it, for example, that governments suppress scientific evidence if it does not support national positions.

Key Points

- For states, international collaboration is never an end in itself. Collaboration is always purposive, the aim being to solve some problems or realize some gains which states could likely not have achieved, left to their own devices.

- Global climate change falls into the sub-class of global environmental problems of long time scale where uncertainty is high, where scientific consensus about causes and consequences evolves slowly, where trade-offs are potentially painful in a socioeconomic sense, and where political inertia ("do nothing, or at least do very little") is great.

- Regional cooperation works to resolve environmental problems only when the scientific questions are of high priority for advanced players and when no other means of attacking them exists.

- Even when regional cooperation "works," one should expect an inevitable rivalry between the regional level and the bilateral level, because the costs of interdependence of the latter are always less.

- The EEZ regime limits the role of regional fisheries commissions, so that most have lost their former management responsibilities to the coastal states. The commissions can only be effective when their activities are harmonized with the objectives and policies of specific states.

- With respect to climate change, the following four lessons can be learned from past experience with fisheries institutions: the nature of the decision system is critical; the divisibility of payoffs is critical; regulation requires monitoring and enforcement; and the role of science is crucial.

Scarcity, Property Allocation, and Climate Change

Jean-Paul Troadec

Before a response to a problem of overuse of a resource can be proposed, its true (as opposed to apparent) cause(s) must be identified. Thus, it is necessary to understand the dynamics of overfishing, and to identify, among the variables involved, those upon which action can be undertaken.

The gap between world supply and demand for marine living resources is widening, even though total catch has been increasing at a slower rate since the early 1970s. As most conventional stocks are fully exploited, there are few possibilities for sustainable intensification of exploitation. Aquaculture can provide a solution but, in the foreseeable future, only for certain commodities and markets.

To cope effectively and efficiently with changes in resource abundance and distribution, instruments to adjust fishing rate and capacity to stock productivity are needed. As evidenced by the situation of overcapacity, overfishing, ineffective enforcement of regulations and conflicts among users within and between uses prevailing in a majority of domestic and international fisheries and uses of aquatic ecosystems, those instruments do not match the new situation of resource scarcity. Schematically, such adjustments are effected through three basic mechanisms: (i) social structure of producers, such as in traditional communities; (ii) application of norms and controls by central governments and administrations; and (iii) market mechanisms. When allocation decisions must be taken to share scarce resources, those mechanisms require clear property regimes in order to function efficiently.

The change in the ocean regime has been an initial, but partial, move in that direction. Presently, there are three basic property regimes in the oceans: (i) open access in the high seas (like the pre-EEZ regime); (ii) national co-ownership of stocks straddling several EEZs; and (iii) national ownership of stocks distributed entirely within single EEZs. Even in the latter case, few countries have clarified as yet the access regime for their domestic fishermen. In many countries, human inputs (capital and manpower) are allocated through economic mechanisms while access rights are regulated through administrative decisions. Since administrative mechanisms do not integrate properly the value taken by access rights when the resources are binding, resources tend to be used as if they were unlimited. Capital and manpower resources are invested in excess, resulting in overfishing. The increases in demand (by increasing the prices of fish) and technological innovations (by decreasing the unit cost of fishing) further the trend. Thus, the problem is not so much one of resource scarcity, but of institutions that are inadequate to cope with the new situation of resource scarcity.

The solution to this issue has two facets:

i) a physical one: owing to the jointness and geographical dimensions of fish stocks, resources can seldom be allotted directly among fishing companies and fishermen; quantitative rights have to be expressed through surrogates (i.e., catch quotas or licenses), and allocated at a level higher than the producers; these are technical arguments in favor of a distinction between resource ownership and exploitation;

ii) an institutional one: property regimes and allocation mechanisms need to be clarified so that, in the production function, resource, capital and manpower inputs could be used in proportion to their relative scarcity.

Such clarification would also be beneficial for balancing conflicting uses of aquatic environments. It would facilitate the development of aquaculture in open waters, since investments depend on the provision of adequate guarantees on farming sites and stocks.

Climate change will have a number of effects on fisheries. The distribution and abundance of individual fish stocks are likely to change, possibly more than will the overall production in a given EEZ or region. However, the directions and magnitudes of individual changes are not yet known. Marine ecology suggests that the spatial scale of change is likely to be important: within a species, populations may either appear or disappear at the boundaries of the distribution of the species. Examples already exist of a variety of potential impacts of climate change on fisheries. They provide analogies, when trying to determine how present fisheries institutions and bodies might cope with future problems. However, while new research will provide better information on the reaction of such structures and mechanisms, the issue of adjusting the institutions to the new conditions will remain.

Presently, fisheries are characterized by their lack of flexibility. Under current management mechanisms, fishermen and administrators tend to stick to the status quo, even when it is markedly inadequate on all grounds and for all parties. Fisheries are, therefore, particularly vulnerable to changes in abundance and distribution of particular fish stocks. By enhancing the mobility of people and vessels with respect to particular fish stocks, institutional innovations can improve the capacity of fleets to adjust to climatic changes.

For large-scale fisheries, with well-established tradition of market economy, economic mechanisms are adequate for allocating individual fixed-term quantitative rights. Provided co-owning countries can agree on how they will share their resources and the owner surplus, recourse to market mechanisms for allocating fishing rights could in the same way improve their capacity to cope with changes in the abundance of shared stocks. Since, in their case, the resource property regime has been clarified, introduction of such mechanisms is likely to be easier for foreign fleets fishing in other countries' EEZs. However, everywhere the development of appropriate institutions is likely to encounter strong opposition. Thus, progress will be slow. Where such innovations are politically premature, the levy of fees on catches or on vessels could partly reduce overcapacity and overfishing. On the other hand, where administrative rationing continues to be used, administrations could be led, owing to the deficiencies of administrative mechanisms, to ossify further the fisheries by, for example, restricting the exploitation of specific stocks in specific locations to specific fleets. In high seas fisheries, owing to the lack of clarification between resource ownership and exploitation, little progress in the effectiveness of fisheries management is to be expected.

In most small-scale fisheries, especially in developing countries, the adoption of market mechanisms for allocating individual use rights would be premature. The allocation for indefinite periods of time of collective exclusive rights of fishing within discrete sea territories to individual communities can provide transitory solutions. Such schemes, by securing guarantee of access to defined resources, would improve the capacity of small-scale coastal communities to cope with climate changes, while facilitating their gradual integration into national and global economies.

Thus, institutional deficiencies are particularly marked with respect to: (i) mechanisms required for allocating exclusive fishing rights to individual, or groups of, fishermen, and (ii) property regime for resources distributed outside areas of national jurisdiction (i.e., at the supranational level).

Institutional innovations and policy formulation are separate issues. Prior adoption of adequate institutions conditions the implementation of important political options.

On the other hand, with appropriate institutions, a significant share of the management task could be decentralized, thus lessening the need for central decisions.

Troadec devised the following grid to analyze the capacity to cope with climate-related changes in fish stock abundance and to illustrate succinctly the capacity of nations and regional organizations to cope with climate-related changes in fish stock abundance. It is an attempt to summarize the chances of success or failure in coping with changes in distribution or abundance of marine living resources. It is meant to be an indicative presentation of observations, not an exhaustive one.

A grid to analyze the capacity to cope with climate-related changes in fish stock abundance

			A	B	C	D	E
	Implications of stock distribution on resource property regime[1]		Stock entirely within one EEZ (sole national ownership)		Entirely stock within more than one EEZ (national co-ownership)		High seas (predominantly) (open and free access)
	Institutions in force for regulating access[2]		Efficient resource access allocation institutions	Inefficient resource access allocation institutions	Multilateral long-term sharing agreement and efficient resource access allocation mechanism	No multilateral long-term sharing agreement and/or inefficient national resource allocation mechanism	Property regime precludes adoption of efficient resource allocation mechanism
1	Changes in resource abundance[3]	Increase in abundance up to the apparition of a new population[4]	Classical situation of fishery development; no explicit international dimension; no explicit political issue;	Similar to A1; lack of sharing capacity masked by the expansion opportunity;	Still similar to A1; international dimension of little concern (expansion opportunity plus sharing capacity);	Similar to B1; lack of multilateral sharing agreement and inefficient allocation capacity masked by expansion opportunity;	Similar to D1; inadequacy of property regime and lack of allocation capacity masked by expansion opportunity;
2		Decrease in abundance up to the collapse of an existing population[5]	Economic, social and political issues minimized by clarity of property regime and efficient resource allocation mechanism (dissociation between resource management and fishing company management - higher mobility of W(labor) and K(capital) relative to R(Resource); no major international dimension;	Economic, social and political issues exacerbated by unclear national property regime and deficiency in resource allocation mechanism (low mobility of W+K relative to R); no major international dimension;	Similarities with A2; potential economic, social and political issues and conflicts reduced by multilateral long term sharing agreement and efficient access allocation mechanism; possibility of conflict between countries depending on the terms of their resource sharing agreement; may require negotiations for amending agreement;	Situation worse then B2; economic, social and political issues exacerbated by: - lack of long-term sharing agreement among countries owning the stock; - deficiency in national resource allocation mechanisms; sharing claims expressed through exploitation; positive aspects include explicitly-limited co-ownership and potential concern of co-owners for resource conservation;	The worst situation; similar to D2, with no positive aspects(unlimited number of parties; no distinction between ownership and exploitation; possession claims expressed through participation in exploitation; low mobility between W+K and R); impossibility to adopt efficient use right allocation mechanism as long as, property regime is not clarified;

			A	B	C	D	E
3	Today's role of regional body		- no primary function; - secondary functions: . diffusion of information on global change potential effects on fisheries, on fishery development approaches; . cooperation in innovative research;	- similar to A3; - diffusion of information, especially on institutional innovations required to upgrade efficiency of fisheries and effectiveness of fishery; - cooperation in research; - management more important than in A3 case;	- major duties of regional bodies: . integrating information on stock potential yields; . reaching agreement on overall fishing regime; . negotiating and revising multilateral long-term sharing agreement; . eventually, reaching agreement on common fishing rights allocation mechanism: - a splitting of traditional regional fisheries bodies possible since body membership change with each stock according to its distribution relatively to EEZ boundaries;	duties of regional body similar to those in the C3 case, but efficiency significantly reduced by absence of long-term sharing agreement(co-ownership) and lack of efficient national (possibly commonly-shared) resource access allocation mechanism;	situation similar to D3, with little prospect for improvement as long as property regime is not clarified, since such clarification conditions adoption of efficient resource access allocation mechanism, and enhanced mobility of W+K relative to R(distinction between resource ownership and exploitation/use);
4	Examples of regional bodies that have coped		most regional fisheries bodies disseminate information; few fisheries bodies, mainly concentrated in the Northern hemisphere(e.g., ICES),harmonize effectively national scientific programs, provide fora for confronting ideas and assessing scientific findings;		Pribiloff fur seal commission; three causes can be put forward to explain its performances in fishery management: . clear property regime(exclusion); . economic mechanism for allocating individual quantitative use rights(distinction between resource property and exploitation, as well as between the owner resource rent and the producer normal profit); . the ashore exploitation of the stock which facilitates control;	- very few examples of satisfactory management effectiveness: . ICES/EEC: temporary banning of fishing after North Sea stock collapse; however, owing to the lack of knowledge on the relative roles of natural and man-made factors in this collapse, measures are likely to be significantly sub-optimal(underestimation of non-fishing factors); . International Convention for Pacific Salmon Fishery: Canada-US agreement; - no example of achievement of fishery efficiency;	- no example of adequate effectiveness of international fisheries bodies, except under external(to the fishery sector) pressures(e.g.temporary banning of whaling adopted under the pressure of ecologists); - however, considering that high seas resources are considerably smaller than those within EEZs and that at world level, tuna fisheries(the major resource) are presently constrained more by demand(market) than by supply (overfishing), domestic/EEZ fisheries have the priority for adjusting institutions to resource scarcity;

			A	B	C	D	E
5	Examples of regional bodies that have not coped		- progress in the ability to cope with decrease in fish stock abundance largely depends on initiatives taken at national level to adopt institutional innovations that are needed to adjust fishing capacities to the productivity of resources that already are scarce; the issue of sharing potential yield among co-owners would be greatly facilitated by distinguishing the resource rent(accruing to the owners) and the normal profit of producers, i.e., by a clarification of the property regime and the adoption of mechanisms appropriate to allocate limited yields; however, attention given to those technical aspects of institutional innovations needed to upgrade fisheries efficiency and management effectiveness *in domestic fisheries* is globally not commensurate with the currently poor contribution of fishery sectors to national economies and the concentration of resources within EEZs(relative to those in the high seas); - so far, attention devoted by fishery biologists to effects on fisheries of global change has been globally limited, especially in tropical regions; - major reasons of such limited interest include more immediate(operational) priorities; lack of innovative investigations in fishery research institutions of member countries; the splitting between regional bodies responsible for fisheries and aquaculture on the one hand, and those covering environment protection on the other hand, whereas, with the intensification and diversification of uses, particular uses cannot be properly analyzed and optimized in isolation from other uses; with significative differences among regions, national innovative research efforts are globally modest in South countries;				- little progress in the effectiveness of high sea stocks management to be expected as long as property regimes are not clarified; - progress in this respect can be expected to be slow considering that: . high seas resources are small relatively to shelf ones; . on a global scale, the major one(tuna) is still not scarce; . in most cases, institutions are adjusted to new conditions only when the pressure to do so is strong.

1 - More categories than those listed in the table are in force - notably, stocks straddling marginally outside the 200 miles, stocks more equally distributed between several EEZs and in the high seas. However, in terms of property regimes, those are clusters of the three basic property regimes analyzed in the table - namely, single national-ownership, multinational co-ownership(e.g. closed seas such as the North Sea), and open access.

2 - Three basic institutions are relevant here: the international property regime(EEZ/high seas), the national property regime(clarified/unclarified; state/collective/individual), and the mechanism for allocating individual fishing rights(market/command(administrative decisions)/social structure(traditional fisheries).

3 - The basic component in the organization of the marine life is the population. Spawning areas of shelf stocks are attached to certain hydrological structures themselves associated to certain bottom features(topography, deposits,...), consequently, the location of shelf populations is geographically stable. This geographic stability is less marked for oceanic populations. Two major effects, both either positive or negative, can be expected from long-term changes in the hydro-climates (i) the appearance or disappearance of populations, without changes in the geographic distribution of discrete locations suitable for the occurrence of shelf populations and (ii) changes in abundance of existing stocks. Appearance/disappearance of populations can be treated as limits in change in population abundance.

4 - A major conclusion can be drawn immediately from this row: whatever the population geographic distribution and the institutions in force for regulating access to the resource, societies will have no difficulty in coping with unexpected bounty.

5 - The preliminary conclusions which could be drawn from this row are: (i) the capacity of societies to face declines in stock abundance or occurrence depends on institutions (property regime and allocation mechanism) in force, first at national level(since the bulk of the resources are distributed within the EEZs), then at international level(high seas), to regulate access to resources that are already scarce; (ii) in this respect, economic mechanisms enhance mobility of labor and capital relative to the resource, but adoption of appropriate market mechanisms depends on the clarification of property regimes; (iii) regional organizations remain necessary for the management of shared stocks, but their effectiveness would be significantly enhanced by clarification of the property regime among co-owning states and the adoption of market mechanisms for allocating quantitative, time-finite, individual use rights to producers; (iv) still,every country and every fishery has not the same capacity to adopt the institutional innovations that would upgrade the capacity to adjust to marked declines in stock abundance.

Key Points

- In many fisheries, human inputs (capital and manpower) are allocated through economic mechanisms, while access rights are regulated through administrative decisions. When allocation decisions must be taken to share scarce resources, those mechanisms require clear property regimes in order to function efficiently. Thus, the problem with fisheries is not so much one of resource scarcity, but of institutions that are inadequate to cope with the new situation of resource scarcity.

- Currently, the insufficient clarification of property regime for stocks distributed outside areas of national jurisdiction, and of mechanisms for allocating use rights to domestic fishermen, are critical weaknesses for adjusting to future changes in fish stock abundance and distribution.

- Examples already exist of a variety of potential impacts of climate change on fisheries. They provide analogies, when trying to determine how present fisheries institutions and bodies might cope with future problems. However, while new research will provide better information on the reaction to change of such structures and mechanisms, the issue of adjusting the institutions to the new conditions will remain.

- By enhancing the mobility of fishermen and vessels with respect to particular fish stocks, the capacity of fleets to adjust to climatic change depends critically on the adoption of institutional innovations.

REGIONAL ORGANIZATIONS

The Organization of American States (OAS) and Issues of Environment and Development

Newton V. Cordeiro

Introduction

Among regional organizations, the OAS is unique because it is the only one that covers an entire hemisphere, from the Arctic to the Antarctic. Within the 35 nations of the OAS, more than 100 cultures exist.

The Americas account for nearly 13 percent of the world's population and 28 percent of its land surface. They cross all the earth's climatic zones and, unlike the continents of the eastern hemisphere, are linked by ocean currents – the eastern boundary currents – along their western coasts. They contain an extraordinary biodiversity: more than 51.8 percent of the world's protected areas and 69.8 percent of the biosphere reserves recognized by UNESCO are located in the Americas. They have three of the largest rivers in the world (the Amazon, the Mississippi, and the Plata) which drain most of their land area. Within the hemisphere, South America has the largest volume of freshwater flow, amounting to twice the runoff of all the other continents combined.

The Amazon rainforest, which spreads over the boundaries of eight South American countries, is suggested to play a major role in the global climate system and is, therefore, implicated in the global warming issue. Also, the OAS member states include some of the world's major producers of carbon dioxide, in both absolute and per capita terms. The impacts of climate change are likely to include all possibilities; the Americas include regions which are likely to be among the most as well as the least affected by global warming.

Since its creation, the OAS has understood development as an integrated process where physical, biological, cultural, social, economic and political variables continually interact. A number of articles in the OAS Charter address the issues of development and environment and the institutional and operational structure of the organization reflects these concepts. Concerns over environment and development have always been present in the philosophy and programs of the OAS. An early example of importance is the 1940 *Convention on Nature Protection and Wildlife Preservation in the Western Hemisphere (the "Western Hemisphere Convention").*

Virtually all activities of the OAS treat the many facets of environmental management. This is especially true of the technical cooperation activities of the organization, and it is now also gaining importance in the legal and political forums of the organization.

The most recent OAS initiative occurred at the twenty-first regular session of the General Assembly, held in Santiago, Chile in June 1991, which established an Environmental Commission within the OAS Permanent Council. The *Inter-American Program of Action for Environmental Protection* is a product of the work of this commission.

In order to assist the member states in their actions to protect and improve their environmental quality, the following measures for regional cooperation are included in the Inter-American Program of Action for OAS implementation:

- To promote cooperation within the hemisphere with regard to global efforts; to treat the effects of climate change; to reduce substances that deplete the ozone layer under the Montreal Protocol; and to examine possible joint hemispheric actions that would broaden and strengthen cooperation in the member countries most severely affected by depletion of the ozone layer.

- To provide technical support and advisory services to member states that would help conclude agreements and projects for environmental protection and improvement, and that would draw on funds that may become available through reduction of external debt.

- To provide advisory services and technical support to member states to help prevent, reduce, or avoid adverse effects from transboundary pollution – including that caused by transboundary movement or discharge of hazardous wastes.

- To promote coordination of policies which look toward conservation of vulnerable resources and ecosystems, including the hemisphere-wide projects for the creation and management of parks and reserves.

The OAS General Secretariat and issues of environment and development

In addition to the Committee on the Environment and the Inter-American Program of Action for Environmental Protection, the OAS Inter-American Economic and Social Council (CIES) and the Inter-American Council for Education, Science and Culture (CIECC) have been actively working in environmental issues for a considerable period of time.

The Department of Scientific and Technological Affairs of CIECC has been working, for the last 25 years, with universities, professionals and scientists of all member countries, in a comprehensive spectrum of issues and disciplines, which include the evaluation of ocean resources, limnological studies, preservation of coastal and terrestrial environments, large marine ecosystems and dynamic processes, and sectoral planning and scientific policymaking applied to the environment.

Large-scale environmental impact studies have included the effects of the El Niño phenomenon on the fisheries and climate of the southeast Pacific Ocean, and the impact of pollution from fossil hydrocarbons on the Caribbean Sea and the tourism economy. Other multinational studies covered large regional ecosystems, such as the arid and semiarid zones of the Andes, Mexico, and Brazil (e.g., Mato Grosso wetlands). In 1992, the Department of Scientific and Technological Affairs became the repository for the Agreement for the Inter-American Institute for Global Change Research, a center for the Western Hemisphere sponsored by the US Government with participation of several countries of Latin America and the Caribbean.

The Executive Secretariat for Economic and Social Affairs of the OAS, through the Department of Regional Development and Environment (DRDE), makes contributions to development planning and to the coordination and execution of technical cooperation projects, and assists governments in the execution of studies at the profile, pre-feasibility and feasibility levels.

This department has undertaken numerous major studies in virtually all member states of Latin America and the Caribbean. Close to 50 percent of the land area of Latin America and the Caribbean has been studied during the department's nearly 30 years of existence. These studies include the entire Plata River Basin, most of Central America, a large portion of the northern Andean zone, and large sections of central and northeast Brazil. At the request of the countries who have signed the Treaty of

Amazonian Cooperation, current development studies will cover much of the Amazon region within the next few years.

The department emphasizes integrated studies, especially for border areas which involve more than one country. Six such studies are currently under way, and activities of this nature appear to be expanding. Another important area of environmental management planning focuses on integrated development of river basins, some of which are shared by more than one country, and now specifically look toward reducing transboundary impacts. Based on the early experience of the department, such projects provide a practical framework for discussion of conservation as a vehicle for development; they facilitate communication within the region on conservation issues and can provide a basis for common approaches to management and control of natural resources along common borders. Pilot projects, where conservation constitutes a major component of development efforts, are also conducted. More important, multinational participation strengthens cooperation among national governments and can lead to multinational agreements in resource management.

The department has also provided technical assistance, training, and technology transfer to OAS member states in natural hazards management since 1983. The DRDE has worked throughout the 1980s in climate-related hazard assessments for use in defining sustainable economic development projects. For example, the DRDE has performed desertification hazards studies in Paraguay (Chaco Region, 1984) and Brazil (São Francisco River Basin, 1987), both for use in integrated development projects. In addition, flood hazard assessments were carried out in Paraguay (Chaco Region, 1984), Honduras (Department of Atlántida and Islas de la Bahía, 1985), Bolivia (Mamoré and Parapetí River Basins, 1987), and Brazil (State of Alagoas, 1989–90), also as part of development projects. An agreement for technical cooperation between the US National Oceanic and Atmospheric Administration (NOAA) and the OAS was signed in 1989, and has recently been expanded to include natural and human-induced climate impact assessment activities, in order to relate emerging climate change issues in a broader environmental management perspective.

OAS multilateral technical cooperation has played an important role in institutional capacity building, training and data management, as well as in the transfer of relevant technology. The OAS uses its experience and comparative advantages in providing technical cooperation in many of the above-mentioned technical fields where the demand for assistance is growing rapidly. One of the things OAS member states need most is a regional system to monitor and report the degree of compliance by states with treaties to which they are parties. If this can be done at the regional level, individual national governments may be less likely to exert disproportionate influence over other OAS members. A regional approach to environmental problems also attracts more technological and financial assistance, when and where necessary, than individual member countries can secure individually.

All technical assistance work is covered by cooperative agreements with the governments of member states. In addition, the OAS can and often does enter into agreements with other multilateral institutions, governmental and non-governmental organizations in order to provide services to member states. Recent work under such agreements includes projects with the Inter-American Development Bank (IDB), the United Nations Development Program (UNDP), the United Nations Environment Programme (UNEP), the Japanese Development Agency (JICA), the US Agency for International Development (USAID), the Canadian International Development Agency (CIDA), the Finnish International Development Agency (FINNIDA), Conservation International, and the Nature Conservancy.

Conclusions

The OAS, as a regional political organization where all 35 independent countries of the hemisphere are members, possesses a comparative advantage in that it brings together both developed and developing countries and provides a regional forum where nations of the north and south have agreed to discuss and seek solutions to the issues of environment and development. Perhaps this political reality reflects the physical and human characteristics of the hemisphere: diversity can lead to complementarity and common solutions.

It is this unique quality that makes it possible for two of the most highly developed countries of the world to unite with the governments of Latin America and the Caribbean and work for the benefit of all the people of the hemisphere. It also presents an opportunity to address the interconnected problems of development and environment through a strategy of joint action. Through the Inter-American system, the OAS can play a significant role in the hemisphere in the implementation of the decisions adopted globally at the United Nations Conference on Environment and Development held in Brazil, in June 1992.

Key Points

- The OAS is unique among regional organizations because it is the only one that covers an entire hemisphere, from the Arctic to the Antarctic. Within the hemisphere, South America has the largest volume of freshwater flow, amounting to twice the runoff of all the other continents combined.

- OAS member states include some of the world's major producers of carbon dioxide, in both absolute and per capita terms. Also, the Amazon rainforest plays a major role in the global climate system. The Americas include regions which are likely to be among the most as well as the least affected by global warming.

- The OAS possesses a comparative advantage in that it brings together 35 developed and developing countries and provides a regional forum where nations of the north and south can seek solutions to the issues of environment and development.

Regional Organizations and Environmental Change

Sam Tewungwa

Over the past few decades, environmental changes such as acid rain, ozone layer depletion, and global warming of the atmosphere have attracted worldwide attention and, as a result, have assumed a high priority on the international agenda. No doubt these issues will continue to receive political attention in the foreseeable future, because of the profound effects they could have on the economic and social welfare of human society. These issues will also retain the attention of the scientific community because their physical and chemical aspects are not yet adequately understood. Society is now contemplating measures to reduce the pace of the evolution of these environmental problems, as well as develop measures designed to prevent, mitigate, or adapt to the changes even before their adverse impacts become manifest.

Environmental changes have consequences that often cross national boundaries. For example, no nation, however rich or powerful, can single-handedly solve the problem of stratospheric ozone depletion. A similar situation exists with respect to global warming. Measures to slow down ozone depletion or global warming are not without costs, which vary from one country to the next. Thus, to implement such measures calls for the pooling together of the world's financial and human resources.

What role can regional organizations play in the combat against global warming as well as other forms of environmental degradation? To put it another way, is there a way to increase the involvement regional organizations in this combat? One way to answer these questions is to assess the manner in which regional organizations can contribute to dealing with the regional consequences of global environmental change.

It is unrealistic to expect each one of a group of, say, four neighboring developing countries to establish institutions to handle separately each regional environmental issue. It would be economically more efficient in terms of human and financial resources for each country to pool existing resources and then encourage each country in the region to focus on a specific aspect of the problem and share the results with other states in the region.

How have regional organizations performed in practice, and how and why were they established?

At the time of its demise in the mid-1970s, the East African Community was one of the oldest regional organizations in Africa. The Community had played a very significant role in the economic, cultural and social affairs of the peoples of Kenya, Tanzania and Uganda. The primary objective of the British in starting this cooperative endeavor, with the construction of the "Uganda Railway" in 1896 from Mombasa to Kisumu at the shores of Lake Victoria, was to exercise more control over their new dependencies; Kenya, Uganda and Zanzibar. In addition, the British also wanted to exercise control over the headwaters of the Nile before the Germans, who at the time were in Tanganyika, could do so.

Over the years, and operating under different names – East African High Commission or East African Common Services Organization, depending on the political circumstances in the region – the East African Community grew into a massive cooperative endeavor employing tens of thousands of individuals and providing a wide range of services to the peoples of East Africa. Until its breakup in mid-1977, the Community was responsible for the administration and management of rail, air, marine and lake transport, including cargo handling and posts and telecommunications on behalf of the Governments of Kenya, Tanzania and Uganda. It also provided meteorological as well as air traffic control services and managed several training establishments. Some of the research institutions developed and managed by the East

African Community included those in agriculture and forestry, marine and freshwater fisheries, industrial research, medical and veterinary research including pesticides.

The Community thus had the potential to contribute significantly to the management of the environment in a regional context. Was that potential fully exploited? An examination of the trends of energy consumption in East Africa from the 1950s to the early 1970s, when development activities were at their peak in the region, may provide an answer to this question.

It is estimated that for 1961–75, the three partners in the East African Community emitted a total of 22 million metric tons of carbon into the atmosphere from the consumption of commercial fossil fuels. The annual increase of these emissions was 5%. Further, during the period 1971–80, the three are estimated to have used 740 million cubic meters of fuelwood as a source of non-commercial energy, in addition to a large quantity of charcoal, bagasse (sugar cane residue used as a fuel), and animal waste. These figures may sound small compared to global figures; however, for a region with inadequate financial resources and pressing environmental problems (including serious land degradation), all efforts ought to have been taken to reduce them even further, through cooperative energy management programs.

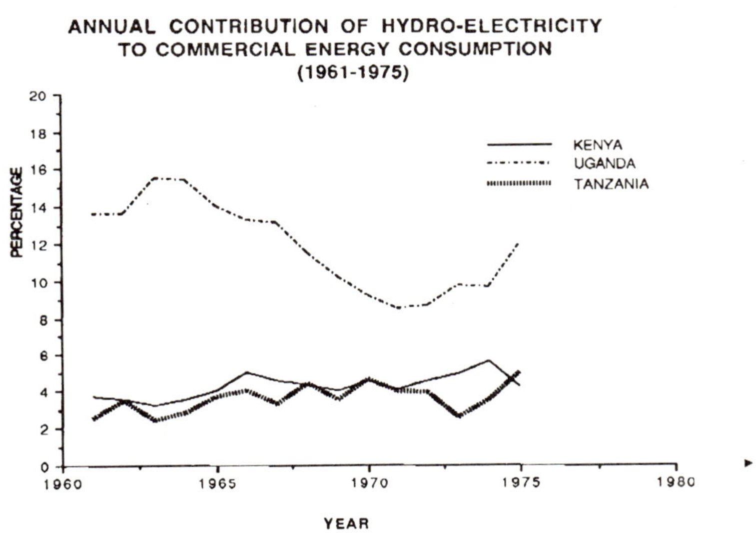

Fig. 1 Annual contribution of hydroelectricity to commercial energy consumption (1961–75).

Hydroelectricity is a cleaner and cheaper form of energy than fossil fuels and much more environmentally friendly than fuelwood and charcoal. As shown in the figure above, this form of energy only contributed an average of 4.2% in Kenya, 3.5% in Tanzania, and 11.9% in Uganda to annual commercial energy consumption during the 15-year period 1961–75.

As of 1961, the installed capacities for hydroelectricity generation was 26 megawatts for Kenya, 20 and 121 for Tanzania and Uganda, respectively. By 1975, however, Kenya had raised it capacity to 166 and Tanzania to 90 megawatts. Uganda's capacity had only risen to 150 megawatts by that time.

As a region, East Africa has great potential for hydroelectricity generation. Although the installed capacity on the Nile (within Uganda) has ranged only between 120 to 150 megawatts for almost half a century, the system is known to have a capacity of 2 gigawatts (2000 megawatts). Viewed in the context of the trends in commercial energy consumption mentioned above, the lack of drive to tap this potential raises some economic and environmental issues:

— Why, in a region with such great possibilities to generate hydropower, should the contribution of this type of energy to commercial energy consumption be so low, particularly in Kenya and Tanzania?

— Was the *individual* massive investment by Kenya and Tanzania to raise their respective production capacity by over 500% and 350% between 1961 and 1975 unavoidable?

Failure to cooperate in the management and utilization of energy resources in East Africa at the commencement of active economic investments is a tale of lost opportunities. The Owen Falls Dam at the source of the Nile, commissioned during 1954, and the grid established for the distribution of hydroelectric power, should have been undertaken as joint, cooperative commercial endeavors between Kenya, Tanzania, and Uganda. If it had been created, an "East African Power and Lighting Company" would have survived the collapse of the community like the East African Development Bank did. This would have saved the region billions of dollars, both in individual development of hydropower generation capacity and in the importation of fossil fuels, in particular considering the tenfold increases in oil prices during the early 1970s. It would have also reduced heavy reliance on fuelwood and charcoal as sources of energy and reduced consumption of fossil fuels, and thereby led to a healthier environment.

While negotiations were under way for the division of the assets and liabilities of the defunct East African Community, a new regional organization, the Southern African Development Coordination Conference (SADCC), was born on 1 April 1980, in Lusaka, Zambia. The original membership of SADCC consisted of Angola, Botswana, Lesotho, Malawi, Mozambique, Swaziland, Tanzania, Zambia and Zimbabwe. It has been joined by a newly independent Namibia. Recall that Tanzania had also been a member of the East African Community. In fact, some of the current SADCC members, notably Botswana and Zambia, had made overtures to join the East African Community prior to its demise. No doubt, with the collapse of one regional organization, the Community had accelerated the formation of another – SADCC.

Established with the main objective of achieving economic liberation and reducing economic dependence on the white-ruled Republic of South Africa, SADCC operates through a series of sectoral commissions, each of which is coordinated by one member state. Some of these sectoral commissions and their coordinators are shown in the following:

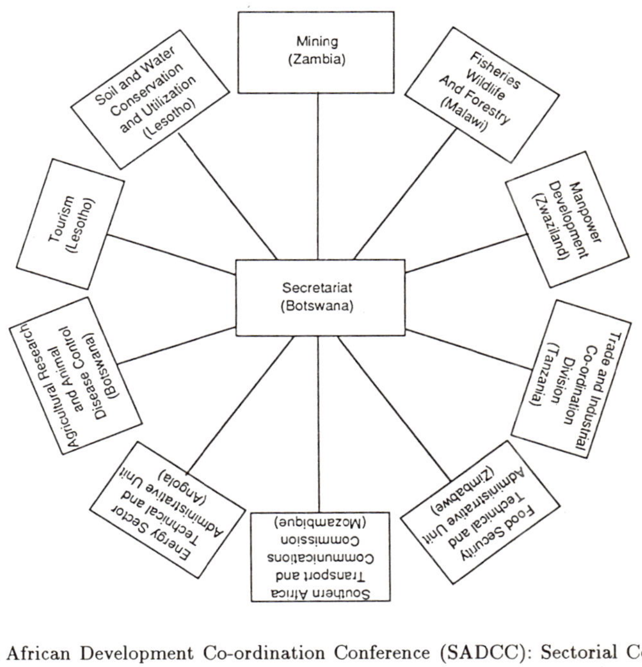

Fig. 2 Southern African Development Co-ordination Conference (SADCC): Sectorial Coordination Offices. From William T. Tow, 1990: *Subregional Security Cooperation in the Third World*. Boulder, CO: Lynne Reinner Publishers.

By the early 1960s, several countries in North and West Africa, and a few in Eastern Africa, had become independent sovereign states. They immediately recognized the need for closer cooperation in order to exert concerted pressure on the colonial powers to grant independence to the majority of the remaining colonies and mandates in Africa, and to coordinate and intensify efforts to eradicate poverty, disease and ignorance from the African continent. With these aspirations, steps were taken to establish an institutional mechanism which could further these objectives; a charter establishing the Organization of African Unity (OAU) was signed in Addis Ababa, Ethiopia, during May 1963.

The OAU has spearheaded the struggle for independence throughout Africa, mediated in several intra-state disputes, and exerted pressure for the abolition of white minority rule and apartheid in South Africa. It has several specialized commissions which focus on specific issues, in particular economic, social, transport and communications, education, science, culture and health, human rights and labor.

Some of Africa's regional organizations are less "global" in scope and were established to address only one issue. For example, the Preferential Trade Area (PTA) which involves more than 12 countries in Eastern and Southern Africa with its headquarters in Lusaka, Zambia, deals only with trade and commercial issues. The Kagera River Basin Organization (KBO), involving Burundi, Tanzania, Uganda and Rwanda, carries out studies on the water balance of the Kagera River. The Undugu Club, comprising Egypt, Sudan, Zaire, Uganda, Rwanda and Burundi, was established in the mid-1980s as a political grouping to foster neighborliness.

The regional organizations mentioned in most of the examples above entailed the negotiation, signature and ratification of a treaty for their establishment. Others, which involve the assessment of a specific issue for a limited period of time, are formed on an ad hoc basis, and their legality is established by the signature of a project document by the countries involved. One example in this category is the Hydrometeorological Survey of the catchments of Lakes Victoria, Kyoga and Mobutu Sese Seko, in which eight riparian countries (Burundi, Rwanda, Zaire, Kenya, Egypt, Sudan, Tanzania and Uganda) have been participating for two and a half decades. The project has established hydrological and meteorological stations and a water quality laboratory. In the long run, the project would formulate a mathematical model for the catchment, which would assist governments in water conservation and the development of water resources for the basin. The project would also develop methods for the optimal regulation of the relevant lakes.

The WMO Tropical Cyclone Project serves as an example of the way in which regional groupings can cope with a climate hazard. Areas around the globe which are adversely affected by tropical storms have organized themselves into five regions, namely the ESCAP/WMO Typhoon Committee, ESCAP/WMO Panel on Tropical Cyclones, the RA IV Hurricane Committee, the RA I Tropical Cyclone Committee for the Southwest Indian Ocean and the RA V Tropical Cyclone Committee for the South Pacific and South East Indian Ocean. The cooperative efforts of the members in each of these regional groupings have led to improved weather observations and data transmission, better forecasts, and the development of early warning systems – thereby minimizing the impacts of tropical storms.

The Convention on Long-Range Transboundary Air Pollution is another excellent example of the manner in which multinational cooperation can be effective in dealing with environmental issues. Since its adoption in 1979, and entering into force in 1983, the Convention has had at least two important protocols annexed to it; one to reduce and control emissions or transboundary fluxes of sulphur compounds (Helsinki Protocol, 1985) and the other to control nitrogen oxides (Sofia Protocol, 1988). According to the UN Economic Commission for Europe, the policies adopted by the Parties to the Convention have led to a reduction of more than 20% in overall sulphur emissions between 1980 and 1990.

The recent pledging conference for drought victims, organized in Geneva in May 1992 by SADCC and the United Nations, provides an excellent example of how effective regional organizations can be in responding to climate-induced disasters. The carefully prepared documentation presented to the conference identified areas of stress in each of the ten SADCC member countries, provided numbers of populations at risk, and assessed in monetary terms the cost of taking remedial action. The developed countries and international organizations present responded by pledging more than one-half-billion dollars (US).

Regional cooperation implies surrender of a certain amount of sovereignty, as collective decisions taken for the good of the whole may not be in the best interests of all of the members all of the time. Furthermore, many national politicians tend to feel that management activities within a regional organization may generate policies which go against their own national policies; they often fear the creation of a supranational body they may not be able to control individually. These political issues, coupled with the fact that regional bodies may also compete with national projects for scarce financial resources, tend to create friction among national members of a regional organization, and may bring about its demise. These are, in part, some of the reasons why the East African Community collapsed, after almost 80 years of existence.

These weaknesses notwithstanding, regional organizations have a significant role to play in coping with problems related to environmental change, in particular those related to climate change and climate variability. Their ability to play a positive role is based on a number of factors, such as the following:

- They can render more efficient service and respond effectively to the problem at hand, as they are to some extent insulated from parochial political influence and maneuvering;
- By pooling human resources drawn from a wide variety of countries, regional organizations tend to be better staffed, offering an optimal way of utilizing meager human resources in the region. Appointments to bureaucratic positions in regional organizations are subjected to less national political interference;
- By avoiding duplication of efforts at the national level in several countries, such organizations can offer services at a much lower cost;
- Corrective measures against environmental damage, developed and implemented at the regional level, can have a greater impact than those taken locally or nationally.

Regional organizations can be an effective institutional tool in combating environmental degradation. Policy makers must make a deliberate effort in encouraging existing regional organizations to include environmental change on their agenda.

Key Points

- No nation, however rich or powerful, can singlehandedly resolve truly global environmental problems such as stratospheric ozone depletion or global warming.
- It would be economically more efficient, in terms of human and financial resources, for countries in a region to pool existing resources and to encourage each country to focus on a specific aspect of a regional environmental problem, sharing the results with other states in the region.
- By pooling human resources drawn from a wide variety of countries, regional organizations tend to be better staffed, offering an optimal way of utilizing meager human resources in the region.

- Regional organization implies a national surrender of a certain amount of sovereignty, as collective decisions taken for the good of the whole may not be perceived to be in the best interests of each of the members all of the time. Furthermore, many national politicians tend to feel that management activities within a regional organization may generate policies which go against their own national policies; they often fear the creation of a supranational body they may not be able to control individually.

The South Asian Association for Regional Cooperation (SAARC) and Climate Change

Saleemul Huq

Saleemul Huq made a brief presentation on some regional organizations in South and Southeast Asia and led a general discussion about the features of regional organizations that make them important (or relevant) to the climate change issue. He then provided a South Asian example (SAARC).

Huq identified three types of regional organizations in South and Southeast Asia:

1) *Regional organizations* defined in terms of the management of specific resources (e.g., river commissions, regional seas organizations); 2) *Economic-political groupings* (e.g., ASEAN, SAARC and, in other regions, EC, OAS); 3) *Regional funding organizations*, usually regional development banks.

1) *Regional organizations* focused on the oversight of specific shared resources can usefully broaden their scope of activities, for example, from navigation to environmental issues. Concern about the possibility of global warming may provide an impetus for regional organizations to consider regional hydrologic issues, especially as the reliability of information about potential climate change on regional water resources increases. There are some successful regional examples. The relatively small Mekong River Commission continued to function, even when one member state was involved in a major conflict with other members of the Commission. Similarly, the Indus River Commission (between India and Pakistan) has managed to function even in times of conflict between its member states.

2) *Economic-political groupings* are composed of states with very mixed objectives, so each grouping must be considered separately. In general, each group must determine its level of interest in an issue before acting on it. For example, in the Association of

Southeast Asian Nations (ASEAN), consensus on economic and political issues was initially difficult to reach. Consensus on environmental issues, however, has been somewhat easier to achieve, and has served as a bonding factor. In the South Asian region, conflicting relationships have long existed between states. Thus, there has been little in the way of meaningful cooperation. Despite the belief that environmental issues might be less controversial, consensus has still been limited, and this lack of consensus extends to the climate change issue as well. In any event, lack of financial resources has precluded action by regional groupings on environmental issues.

3) *Regional funding organizations*, such as regional development banks, are often among the few sources of funds at the regional level. Such organizations are important because they can support the collection, synthesis, and dissemination of scientific information throughout a region. For example, the Asian Development Bank has been very supportive of the environmental activities of its member countries. It has recently supported studies of the possible consequences of global climate change in Southeast Asia, including Vietnam.

A South Asian example

The South Asian Association for Regional Cooperation (SAARC) was formed in 1983 at the initiative of President Ziaur Rahman of Bangladesh. It consists of seven countries: India, Pakistan, Bangladesh, Sri Lanka, Nepal, Bhutan, and the Maldives, all of which lie in the South Asian subcontinent. This grouping of countries is quite heterogeneous; it includes one very large country (India) with over 700 million people, a group of mid-sized countries (Bangladesh, Pakistan, and Sri Lanka), and a group of very small countries (Nepal, Bhutan, and the Maldives). The other characteristic to note is that the entire region has a long history of hostility, with Pakistan having fought several wars with India in the last few decades and Bangladesh having been created by a war of independence from Pakistan, of which it had been a part until 1972.

However, despite their differences, the countries of SAARC do share a number of common interests, features, and history; for example, for nearly 200 years they all were part of the British empire, which has left them with a common legacy in terms of the use of the English language, the educational system, the legal system, and military systems. Each of them is also a member of other, larger international groupings such as the Commonwealth, the Non-aligned Group, the Group of 77, and the United Nations.

SAARC functions mainly through the annual meetings of heads of state of each member country with a very small secretariat based in Kathmandu (Nepal). In recent years there have been attempts to identify and prioritize areas for further strengthening of ties and working together on issues of common interest. Progress, however, has been rather modest, mainly because of the lack of financial resources that would be needed in order to take on any large-scale cooperative enterprise, but also as a result of the history of suspicion between various countries in the region. Such suspicion makes it difficult to arrive at consensus on any issue of major importance.

The work of SAARC has also been hampered by the occasional flare-up of tension and hostility between some member states. For instance, at the moment there is a great deal of tension between India and Pakistan over the issue of Kashmir; India has accused Pakistan of harboring Kashmiri separatists. Similarly, the summit meeting that was to have been held in Sri Lanka a few years ago could not take place because of conflicts between Sri Lanka and India over the Tamil issue.

Environmental activities

One of the areas of common interest and activity is the environment. There have been sporadic attempts to cooperate on environmental issues. The most notable attempt at cooperation was the meeting of environment ministers of the SAARC countries, held in Delhi (India) in early 1992 to coordinate the SAARC input into the preparatory work for UNCED. Although a worthy effort, it did not result in very much concrete coordination of the SAARC countries at the Earth Summit or in its preparation. However, individual countries from SAARC did play a significant role in the UNCED process, as well as in the conference itself. Pakistan, for instance, as Chairman of the Group of 77 (the caucus of developing countries), chaired negotiations with the OECD countries and, thus, played a very significant role in the North–South negotations which took place. Similarly, the Maldives played a very significant role in climate change negotations as a representative of the most vulnerable countries and was also the main driving force behind the organization of the small island countries (the Association of Small Island States, AOSIS), which played a major role in the negotations. At the Earth Summit itself, India, through its Environment Minister and other members of the delegation, played a very prominent part in the debate, particularly with the media on behalf of the countries of the South.

Thus, individual SAARC member countries have played positive roles in the climate change and UNCED process, but not really at an organized or coordinated level.

Future possibilities

The main achievement of SAARC in its first few years has been the very formation and survival of the organization in the face of such a considerable history of regional hostility between its various members. SAARC is now quite well established and all members definitely feel the need to strengthen it further.

It is also quite clear that environment is one of the areas in which there is widespread consensus on the major issues and approaches to be taken. This is particularly true in the area of international environmental issues such as global climate change. However, in order for them to concretize their concern and agreement on the issues into concerted action, both in their own region, as well as in international negotiations, a great deal more concentrated effort and support would be required.

SAARC is still at an early stage in its development. Probably, SAARC's most promising link to the climate issue is their concern regarding the likely increase in hydrologic disasters resulting from climate change. This will require increasing cooperation both within the region and with other regional organizations with experience, as shown by the involvement of the US armed forces in humanitarian disaster relief activities following the 1991 Bay of Bengal cyclone. While such disasters are likely to recur, the prospects of a climate change decades in the future make regional preparedness for potential consequences today even more important.

Key Points

- Concern about the possibility of global warming may provide an impetus for regional organizations to consider regional hydrologic issues, especially as the reliability of information about potential climate change on regional water resources improves.

- In the South Asian region, conflicting relationships have long existed between states. Despite the belief that environmental issues might be less controversial, consensus has still been limited, and this lack of consensus extends to the climate change issue as well. In any event, lack of financial resources has precluded action by regional groupings on environmental issues.

- The main achievement of SAARC in its first few years has been the very formation and survival of the organization in the face of considerable history of regional hostility between its various members. SAARC is now quite well established and all members definitely feel the need to strengthen it further.

The Roles of IGOs in International Environmental Management: Arena or Actor?

Arild Underdal

International organizations can serve at least two major functions in international environmental management: that of being an *arena* for the exchange of information, discussion and decisionmaking, and that of being an *actor* in the policymaking or the policy implementation process. All intergovernmental organizations (IGOs) serve as arenas, but only a subset can qualify as significant actors in their own right.

An *arena* regulates the access of actors to problems (and vice versa), and the access of problems to formal occasions for decisions. Moreover, it defines the official purpose and specifies the rules, location and timing of decision "games." In addition, it may develop a distinct informal "culture," into which participants are more or less effectively socialized. To qualify as *actor*, an organization must (also) provide independent inputs into the policy process, or somehow amplify the outputs of the process. To meet these requirements, an organization must have a minimum of internal coherence (unity), autonomy, resources, and also in fact engage in some relevant external activities. Without a certain minimum of internal coherence, it would not constitute a unit. Without some autonomy in relation to its members and other actors, it would be a mere puppet, the inputs of which should be attributed to its masters rather than to the organization itself. A minimum of political resources is required to become recognized as a relevant (potential) actor in the process. Finally, by definition a real actor must somehow *act*, i.e., undertake some activity in the "game" in focus.

Arenas differ in terms of, inter alia, *rules of access, decision rules* and *rules of procedure*. For example, membership is restricted in some cases to countries that satisfy certain criteria (as is the case with, for example, the Antarctic Treaty System). Other organizations (such as the International Whaling Commission [IWC]) are open to any state that cares to submit a formal application and pay its membership fee.[1] The rules

[1] In fact, the failure to pay membership fees seems *not* to imply automatic exclusion.

of access to particular (executive) bodies within the organization will often be more "restrictive" than criteria for membership in the organization as such. Consensus is the decision rule most frequently subscribed to in IGOs, but a number of organizations have some kind of provisions for decisionmaking by voting (usually qualified majority). Finally, we know that procedural arrangements often differ in several respects, for example, with regard to differentiation into subprocesses (committee work vs. plenary sessions), and the role of the committee or conference chairs in producing (single) negotiating texts.

One major research question generated by these observations becomes: to what extent and how do different rules and arrangements affect the capabilities of organizations-as-arenas to fulfill certain critical functions in the decisionmaking process – including (a) those of providing actors with incentives to adopt and pursue a "constructive," problem-solving approach; (b) providing procedural opportunities for transcending initial constraints (e.g., by coupling or de-coupling issues); and (c) institutional capacity for integrating or aggregating actor preferences.

Organizations – and even specific bodies within organizations – also differ in terms of *actor capacity.* For example, the Commission of the European Community clearly has a much greater capacity to serve as a political actor than does the secretariat of the IWC. Similarly, NATO certainly has a greater capacity to act in European security affairs than the Conference on Security and Cooperation in Europe (CSCE). As indicated above, actor capacity can provisionally be conceived of as a function of internal coherence, the scope and depth of external autonomy (notably vis-à-vis member states), and the amount of political resources at the organization's own disposal. At a more operational level, this leads to research questions such as the following: (a) Which (kinds of) functions do IGOs fulfill in different regimes? (b) What institutional arrangements does the organization have for performing these functions, and how many resources does it have at its disposal? (c) What is the relationship between the actor capacity and actual functions of an IGO on the one hand and its problem-solving "success" on the other?

The distinction between organizations-as-*arenas* and organizations-as-*actors* does not imply a ranking in terms of *importance.* Institutions can shape outcomes as much by coupling actors, problems and occasions for decision, and by determining the rules of the game, as by entering the game as more or less significant actors. Arenas are important in their own right and for *different* reasons. In the familiar terms of systems analysis, we may say that actors influence outcomes by providing inputs (of various kinds) and by amplifying outputs, while arenas are important because they provide opportunities for articulating interests and beliefs and for aggregating preferences, and also because they provide a setting regulating the game itself. Thus, it hardly makes much sense to consider one of these functions as inherently "more important" than another.

More specifically, organizations-as-*arenas* can affect actor behavior and collective decisions in several ways. First of all, by providing an opportunity for discussion and decision, an organization may affect the political agenda of its member states. The mere fact that a conference is about to take place can serve as a stimulus for policymaking at the national level. Moreover, the fact that certain measures are being (or are going to be) discussed, may affect expectations of governments and societies. For example, the fact that restrictions on the use of CFCs were being discussed probably signalled to some producers and consumers that there was some probability (>0) that new restrictions would be forthcoming. Such expectations may in turn have had some (in most cases probably marginal) effect on incentives to look for substitutes, perhaps leading some actors to undertake unilateral adjustments – possibly even before any formal decision to impose restrictions had been made.

Second, international institutions can be important as *frameworks for learning*. For example, the first stages of the negotiations on measures to control anthropogenic sources of global climate change may be seen very much as exercises in the development of consensual ("diagnostic") knowledge that can serve as a basis for the design of possible "cures." In many cases IGOs provide distinctive inputs of their own into that process. But even when they do not, they at least provide a *setting*, "structuring" the games themselves.

Third, organizations can serve as arenas where members can *articulate* interests and beliefs, demands and offers, complaints and support. Moreover, organizations can be arenas where members can have their behavior in a certain issue-area evaluated by others. At least some forms of performance review can provide significant incentives to governments to improve their records, and such instruments may work even in the absence of a formal convention or regulation. In some cases the mere risk of becoming subject to public criticism can lead a government to make some adjustments in its own policies and behavior. Some environmentalists tend to be too quick in dismissing institutional arrangements, such as procedures for performance reviews, as poor substitutes for what they consider to be "the real thing" (substantive regulations).

In fact, procedural arrangements can be important instruments for achieving unilateral adjustments or collective action at a later stage.

Fourth, all organizations have some rules for the *aggregation* of preferences. These rules can affect the behavior of individual actors (e.g., their incentives to pursue accommodative strategies), the formation of coalitions, and of course, the output itself. The decision rule is one important determinant of the amount of influence that individual actors can hope to achieve over outcomes. Moreover, organizations can play an important role in managing or resolving "internal" conflict. Even if no mediation activities are performed by officials or bodies of the organization itself, it can provide a setting in which such activities can be undertaken, and help bring mediative assistance of some other party to bear on the dispute.

Last, but not least, institutions-as-arenas can foster informal, interpersonal networks ("epistemic communities"), and facilitate the development of mutual confidence and a sense of common purpose. Accounts of, among others, UNCLOS III and the negotiations about the ozone regime clearly suggest that working together over an extended period of time can have a number of important side-effects on interpersonal relationships (and although some animosities and tensions can be expected, the net balance is likely to be positive).

Although not exhaustive, this list clearly indicates that even in the role of a "mere" arena an organization can fulfill important functions for its members and possibly for others as well. It seems to me that the significance of such "arena functions" tend to be under-estimated in much of the public debate about the role of IGOs in environmental management. At least in this particular field, most IGOs seem yet to deserve more interest as arenas than as actors.

This is, of course, not to deny that some IGOs (or even specific members within IGOs) do qualify as significant *actors* in efforts at coping with environmental problems. As indicated earlier, organizations can act by providing distinctive inputs and/or by amplifying outputs. On the input side, we may observe that a number of organizations (e.g., ICES, OECD) contribute substantially to the *production and dissemination of (consensual) knowledge*. Anthony Scott[2] argues that this is the most important function that an international organization can possibly fulfill in international decisionmaking processes. Secretariats and conference or committee chairpersons can also serve as

[2] Anthony Scott, 1976: Transfrontier pollution: Are new institutions necessary? In *OECD: Economics of Transfrontier Pollution.* Paris, France: OECD.

political *entrepreneurs*, helping to design solutions that are politically feasible and as substantively effective as "circumstances permit." Similarly, organizations can amplify outputs by, for example, monitoring compliance, carrying out their own projects, or allocating resources to help members implement decisions or achieve agreed goals. Even as a passive arena an organization can add *something* to a collective decision, notably its "seal of legitimacy." Thus, a resolution passed by the UN Security Council would probably command in most countries somewhat higher political "status" than a set of unilateral but substantively equivalent declarations given outside the framework of the UN. And the fact that a (temporary) ban on commercial whaling has been formally adopted by the appropriate intergovernmental authority (IWC) seems to have served, in internal discussions, as the principal argument against the decision taken by the Norwegian government to permit some commercial catch from one particular stock in 1993.

Even though *regional* organizations primarily are instruments for coping with regional problems, they can serve as important arenas and actors also for *global* problems. More specifically, a regional organization can influence global negotiations by facilitating coalition formation and serving as an instrument for defending or promoting the interests of its members, or by helping its members work out a joint program of contributions to a global program. In the context of global environmental problems the latter function may deserve particular interest. To give just one example: agreement within what might be called the OECD area on some "formula" for "internal" distribution of costs involved in controlling anthropogenic sources of greenhouse gases would certainly be one major step towards a global climate change regime. The search for such a formula can be facilitated substantially through the services of a *regional* (or at least non-universal) IGO – preferably one that could provide its own distinctive inputs into the process as well as offer a suitable arena for negotiations. (In this particular case the OECD – a "multiregional" organization – seems to be the major candidate, with the EC as the most important truly regional component).

With respect to the possible actions of organizations in specific circumstances, four caveats must be made:

1) "Success" is a function of the "benignity" of the problem and the problem-solving capacity of the institution(s) and actors involved in coping with it. Some problems are clearly more "benign" than others and hence easier to solve. In considering the accomplishments of an organization it is therefore important to control for

the character of the problem; in dealing with a politically "malign" problem, even a marginal improvement compared to the status quo may be a substantial achievement.

2) Efforts to solve problems may have important side effects. Thus, the initial stages of the UNCED process can be seen as much as an exercise in *search* and *learning* as an effort at negotiating a new regime. Furthermore, international problem-solving efforts can lead to a number of *unilateral* adjustments on the part of societies as well as governments, and the aggregate impact of such adjustments may be as important as the impact of whatever collective measures are adopted.

3) Institutions may themselves be part of the solution. Without questioning the primacy of substance over procedure, the institutional arrangements agreed upon in, for example, UNCED, should not be seen merely as poor substitutes for the "real thing," but also as potentially important instruments for regime development and implementation.

4) Sometimes, however, international regulations may produce outcomes that are *worse* than those that would have occurred had no such regulatory action been taken. The regulations agreed upon by the International Whaling Commission (IWC) in its early stages may be a case in point. Arguably, these regulations failed because – in the absence of a firm basis of consensual knowledge or coercive leadership – total allowable catch (TAC) was set at a level far too high to secure sustainable yield. The failure was exacerbated by the fact that the wrong kind of regulations were used; in the absence of quota allocations, TACs induced a race (sometimes referred to as the "whaling olympics") and substantial overinvestment in whaling fleets (making future reductions in TACs more, rather than less, difficult to achieve).

Finally, it should be added that environmental damage is typically a *side* effect of human activities undertaken for some other perfectly legitimate purpose, such as the production of goods, the transportation of people and goods, etc. This implies that effective environmental policies will have to "penetrate" those activities that cause harm in the first place. Therefore, organizations managing those activities will be, at least potentially, the most effective institutions for coping with international environmental problems. In the European context, environmentalists would be well advised to focus their attention as much on the EC as on specialized regional commissions for environmental protection. Even a modest change in the policies of an organization "in charge" of important policy instruments may achieve more in terms of environmental quality than substantial influence over an institution that has little influence over those activities that determine the state of the environment.

Key Points

- International organizations can serve at least two major functions in international environmental management: that of being an *arena* for the exchange of information, discussion and decisionmaking, and that of being an *actor* in the policymaking or the policy implementation process.

- The distinction between organizations-as-*arenas* and organizations-as-*actors* does not imply a ranking in terms of *importance*. Institutions can shape outcomes as much by coupling actors, problems and occasions for decision, and by determining the rules of the game, as by entering the game as more or less significant actors.

- Even though *regional* organizations primarily are instruments for coping with regional problems, they can serve as important arenas and actors also for *global* problems.

- The best organizations for developing regulations are those whose activities caused the harm in the first place. A focus must be placed on those organizations that have direct links to desired actions and outcomes.

Role of Regional Organizations in the Context of Climate Change

Naresh C. Singh*

Not all international organizations participated to the same degree or with the same interest in the negotiations leading to the Framework Convention on Climate Change. These can generally be divided into the following groups:

- Inter-governmental organizations: OECD, EC, CARICOM, SPF (South Pacific Forum) were active; OAS, NATO, ACP, and African groups were not.

- UN groups from all regions were active.

* Dr. Singh has recently changed his affiliation. His new address is IISD (International Institute for Sustainable Development), 161 Portage Avenue East, Winnipeg, Manitoba, Canada R3B 0Y4.

- Informal groups of nations: G-77 and AOSIS (Alliance of Small Island States) were active. While there was almost no consensus in the G-77, AOSIS was strongly committed and concerned. The G-7 played little role, as they were effectively represented by the OECD.
- Specialized UN agencies (WMO, UNEP, FAO, UNESCO, IOC, UNDP, UN General Assembly) were all involved.
- Regional resource and financial organizations (e.g., regional development banks, fisheries and water basin commissions) were not involved.
- Regional and international NGOs were involved for the first time in major international negotiations.

Prior to the recognition of climate change as an issue for concern, regional organizations were already undertaking a number of activities with respect to climate (e.g., variability and extreme meteorological events):

(a) routine meteorological operations (data collection and analysis, forecasting) and climate research: WMO regional groups, national hydrological and meteorological organizations, etc.;

(b) preparation, response, and mitigation for extreme events (or disasters, e.g., cyclones, droughts);

(c) resource management (agriculture, fisheries, forestry, land use, marine and freshwater resources); although climatic factors were not often given much emphasis.

Regional organizations may be able to play a number of roles relating to climate change, as follows:

- *Framework Convention on Climate Change*: ratification, compliance, development of protocols, and implementation. Implementation could include transfer of technology and financial resources, information exchange, and capacity building.
- *Science*: research on activities to reduce uncertainties regarding climate change.
- *Impacts*: development of regional scenarios and strategies for monitoring, adaptation, and mitigation, to be used also in planning responses at national and smaller scales.

- *Research and action on other related changes*: atmospheric pollution, ocean resource management, and forests.

In general, the actions of regional organizations to date may be described as "bark and no bite." These include: mobilization of information, research, and resources; training and technical assistance; and consensus building (and sometimes coordination, although this is rare even at the national level).

A limited review of regional organizations reveals some general problems: requirement for unanimous decisions, even when the populations of member states differ greatly in size; regional decisions made in spite of lack of national machinery, people, or other resources to implement them; and lack of national commitment to regional agreements.

In some cases, regional organizations may appear to be unsuccessful because their activities are measured against unrealistic expectations, for example: (a) work plans may have to be agreed to by governments; (b) funding may have to come from outside as well as inside the region; and (c) member states must be happy with the implementation of programs; thus, a regional organization sometimes undertakes a few activities in each member state in order to satisfy local political demands.

It is not clear that climate change has special features that make it different from past problems with which regional organizations have dealt. The issues are generally not new, merely more complex. Regional organizations with a more interdisciplinary approach to decisionmaking may be required, as this is not a characteristic of existing regional organizations.

While few existing regional organizations have been explicitly concerned with climate variability or change, many have considered other regional environmental issues. There are now a number of new organizations which focus on climate change (e g., AOSIS). There is a particular need for work on the economic implications of climate change. This might be carried out under the auspices of OECD and GATT.

A. To examine the role, if any, of regional organizations in Climate Change we need to address several fundamental questions. A random few would include the following:

A.1 Are there legitimate concerns with regard to global climate change?

Yes, on the basis of the IPCC science and its uncertainties and the Precautionary Principle, the Ministerial Declaration of 1990 and the Earth Summit Resolutions.

A.2 If so, who are the concerned parties? (states, local groups, UN organizations, regional organizations, researchers, environmental lobbyists, developed countries, developing countries, countries with peculiar ecological endowments such as forests, deserts, mountains, island ecosystems, low lying coasts)?

The really concerned parties are the small low-lying communities of the South Pacific. The concerns of the other parties are predicated on conscience, opportunities for gain and pressure from grassroots and other lobbying groups and consequently political direction.

A.3 What are the parties concerned about? (economic and trade impacts, ecological disasters, disruptions in the hydrological cycle, rainfall distribution, droughts, floods, storm intensity and frequency, security, physical extinction of states, habitats, peoples, cultures, livelihoods, status quo)?

Self-interest, which therefore depends on their circumstances.

A.4 Are the concerned parties prepared to act on or address their concerns and, if so, in what way(s)?

Several countries have shown the will to act but predicated on A.3. Some are prepared to cut back and stabilize CO_2 emissions, contribute to new and additional funding, and provide technological transfers at concessional rates. Others are not. Environmental NGO science and lobbying has been useful.

A.5 Do the concerned parties expect others to act on their behalf and, if so, whom and in what way?

The smaller developing countries are likely to call upon the resources of their regional organizations to assist in capacity building to monitor and comply with the convention. They would probably seek assistance also in some of the key interventions in B.8 (below). They will also seek the assistance of the specialized UN agencies UNEP, WMO, FAO, IOC, etc., perhaps in preference to regional organizations. The developed country parties are unlikely to utilize regional organizations in the near term, excepting of course the European Community.

A.6 Are there parties that consider themselves potential winners and others that consider themselves potential losers, or are all likely to win or to lose?

All parties have been portraying themselves as potential losers.

A.7 What are the major vested interests of the parties, either individually or in groups?

Maintenance of the existing global status quo, economic dominance, development at all costs, maintaining existing life styles.

A.8 What are the chips available to the various players ?

Financial and/or military power, industrial and technological development, exclusive economic zones, energy efficient technologies, market sizes, forests, utilization of polluting technologies.

A.9 Are there "pawns " amidst the bigger players?

Small island and small low-lying coastal states.

A.10 What are the north-north and south-south divisions in addition to north-south divisions?

They are based on the chips in A.8. To assume a sharp north-south divide as the most important one would be seriously misleading.

A.11 What is the relevance of recent global events such as the end of the Cold War, the reunification of Germany, the emergence of the mega-trading blocks, the current recession?

The bargaining chips are economic, trade and development driven rather than based on strategic or military advantages.

B. On the basis of the IPCC science and scientific uncertainties and the provisions of the Climate Change Convention, a second set of issues relevant to the role of regional organizations would arise such as the following:

B.1 Can regional organizations influence their members to sign, ratify and comply with the provisions of the Convention? If so, how?

By demonstrating opportunities for gain where these exist and assisting in technical and administrative preparations and follow-up.

B.2 Can regional organizations act as mediators in achieving compromises and resolving differences in positions of members with respect to the development of protocols and related implementation needs of the Convention?

Yes.

B.3 Is there a role for regional organizations in key areas of implementation of the Convention such as (a) removing scientific uncertainties by promoting relevant scientific research and strengthening research and monitoring capability; (b) facilitating financial flows and resource mobilization; (c) facilitating technology transfer; (d) providing legal and technical support to developing member state parties in their attempts at Convention implementation?

Yes.

B.4 Can regional organizations assist in compliance monitoring of the provisions of the convention?

Unlikely, unless agreed to by the parties to the Convention.

B.5 Are there roles that are regional organizations better placed to play than UN organs and specialized agencies and vice-versa?

Limited. Technical assistance brokerage.

B.6 Are there roles that regional organizations are better placed to play than national organs and agencies?

Resource mobilization and projects development where regional approaches are warranted. Since climate change impacts will largely be determined by local conditions, role of regional organizations will be limited to scenario development and strengthening national capability to respond.

B.7 Can we distinguish some categories of regional organizations and determine some special contributions they are better placed to make?

Economic cooperation, functional cooperation, military space, communications, technical and scientific, professional bodies, NGOs.

B.8 What are some of the key interventions that would be required if the Convention goal is to be realized?

- *stabilization and cut back of CO_2 and other greenhouse gas emissions;*
- *technology development and transfer;*
- *public education and awareness;*
- *information exchange;*
- *provision of new and additional financial resources;*
- *pledge and review process;*
- *insurance mechanisms;*
- *compliance monitoring.*

C. The special case and concerns of small islands and small low-lying coastal states as front line states in an IPCC business-as-usual climate change scenario and the role of regional organizations (AOSIS, CARICOM, SPC, CEHI, SPREP CMO, CIMH, etc.) include:

- Sea level rise, storms, rainfall distribution, temperature change;
- Physical survival;
- Ecological and economic vulnerability;
- Cultural survival;
- Ability to respond (accommodation adaptation, retreat, mitigation).

D. Profiles of as many regional organizations as possible should be compiled.

Key Points

- It is not clear that climate change has special features that make it different from past problems with which regional organizations have dealt. The issues raised are generally not new, merely more complex.
- The actions of regional organizations to date may in general be described as "bark and no bite." These include: mobilization of information, research, and resources; training and technical assistance; consensus building (and sometimes coordination).
- Regional organizations with a more interdisciplinary approach to decisionmaking may be required, as this is not a characteristic of existing regional organizations.

PART IV

Appendix

BIOGRAPHICAL STATEMENTS OF PARTICIPANTS

NATO Advanced Research Workshop

The Role of Regional Organizations in the Context of Climate Change

James P. Bruce

Jim Bruce has worked in the fields of meteorology, water resources and environment. Early in his professional life, he was a weather forecaster, established the flood warning system for the province of Ontario (1955-59), and undertook research and teaching in hydrometeorology. In 1967, he was appointed as the first Director of the Canada Centre for Inland Waters, Burlington, Ontario, and subsequently Director General, Inland Waters Directorate for Canada based in Ottawa. In 1977 he became responsible (Assistant Deputy Minister) for the Environmental Management Service of Environment Canada, comprising the Canadian Forestry Service, the Canadian Wildlife Service, Inland Waters and the Lands Directorate.

In 1980, Jim Bruce became the Assistant Deputy Minister of the Atmospheric Environment Service, Canada's national weather service, responsible for weather forecast and warning services, the Canadian Climate Centre and national air quality monitoring and research. From 1986 to 1989, he served as Director of Technical Cooperation and Acting Deputy Secretary-General of the World Meteorological Organization, Geneva.

In the 1970s he was Canadian co-chair of the Great Lakes Water Quality Board of the International Joint Commission (IJC), chaired the Prairie Provinces Water Board, and the Mackenzie River Basin Board, among others.

In the 1980s he chaired:

- Canada's Federal/Provincial Steering Committee on Acid Rain;
- The Economic Commission for Europe's Working Group, which negotiated the Helsinki Protocol for SO_2 Emission Reductions; and
- The landmark Conference in Villach, Austria (1985) on Greenhouse Gases and Climate Change sponsored by WMO/UNEP/ICSU.

While acting as Deputy Secretary-General of WMO, Jim Bruce assisted the Secretary-General in overseeing the establishment of the Intergovernmental Panel on Climate Change (IPCC), in collaboration with UNEP, and the Global Atmosphere Watch for assessment of chemical changes in the Earth's atmosphere. He had general responsibility for WMO technical cooperation projects, including those for four regional Tropical Cyclone Programs and for a number of agrometeorological and hydrological programs. He has written several books, papers and articles, especially on climate change, environmental management, water resources, disaster mitigation and related topics.

He currently (1992) chairs the Canadian Climate Program Board, the United Nation's Scientific and Technical Committee for the International Decade for Natural

Disaster Reduction and the Scientific Advisory Committee (SAC) to UNEP for the World Climate Impacts Assessment and Response Strategies (WCIRP) component of the World Climate Program. He is also a member of the International Geosphere-Biosphere Program START Committee on Regional Centres for Global Change studies, and undertakes consulting assignments.

Lynton Keith Caldwell

Dr. Caldwell is the Arthur F. Bentley Professor Emeritus of Political Science and Professor of Public and Environmental Affairs at Indiana University, active in environmental and science policy research. His academic degrees are Ph.B. Honors in English (1934) and Ph.D. (1943) University of Chicago, M.A. (1938) Harvard University; and LLD (1977) Western Michigan University. He has served on faculties of the University of Chicago, the University of Oklahoma, Syracuse University, and the University of California at Berkeley, with shorter appointments and lectures at more than 80 other collegiate institutions in the US and abroad. Services to the public include the US Senate, Departments of Commerce, Defense, and Interior, Office of Technology Assessment, National Institutes of Health, United Nations, and UNESCO. Scientific bodies in which he has participated include the National Research Council; National Commission on Materials Policy; Sea Grant Advisory Board; Science Advisory Board of the International Joint Commission; and the Pacific Science Congress. He has served on editorial boards of a number of scientific and professional journals and on the Board of Governors, The Nature Conservancy (1959-1965). He is noted as a principal architect of the National Environmental Policy Act of 1969 and the "inventor" of the environmental impact statement.

He has been Guest Scholar at the Woodrow Wilson International Center, Visiting Fellow at the East-West Center, recipient of the William E. Mosher Award (1964), the Laverne Burchfield Award (1972), and the Marshall E. Dimock Award (1981) of the American Society for Public Administration. In 1989 he received the IAIA-Rose-Hulman Award for Outstanding Achievement in the field of impact assessment and in 1991 was named one of the UNEP Global 500 for distinguished environmental services. He is a Fellow of the American Association for the Advancement of Science; the Royal Society of Arts; and of the International Association of Naturalists; recipient of the Most Noble Order of the Crown of Thailand; member of the National Society of Arts and Letters; US National Academy of Public Administration, Honorary Life Member of the International Association for Impact Assessment, and corresponding member of the National Academy of Law, Cordoba, Argentina. He has published more than 200 articles and monographs and twelve books (translations in 19 languages), of which the more recent are *Science and the National Environmental Policy Act* (1982); *International Environmental Policy* (2nd rev. ed. 1990); *Biocracy: Public Policy and the Life Sciences* (1987); with others, *Perspectives on Ecosystem Management for the Great Lakes* (1988) and *Between Two Worlds: Science, the Environment Movement, and Policy Choice* (1990); and with Kristin Shrader-Frechette, *Policy for Land: Legal and Ethical Relationships* (publication anticipated 1992). His first book, *The Administrative*

Theories of Hamilton and Jefferson (1944), has been twice reprinted, most recently in 1988.

Biographical detail in *Who's Who in America; Who's Who in Science and Engineering; American Men and Women of Science; Dictionary of International Biography; Contemporary Authors;* and *Men of Achievement: International Biographical Centre.*

Newton V. Cordeiro

Since July 1967 Mr. Cordeiro has been a senior specialist in the Department of Regional Development and Environment of the Organization of American States. Previously, 1956–1967, he held staff positions with the National Department of Reclamation (DNOS) in Brazil, where he reached the rank of Director of the Research Division. He holds a degree from the National School of Engineering of The University of Brazil (1956); postgraduate studies in water resources planning and river basin development in Grenoble, France (1957–58), and in-service training in SOGREAH/NEYRPIC, Grenoble, and Electricité de France, Paris. His interest and recent experience are in regional planning, watershed management, border-area integration and development, and the identification and mediation of conflicts in resource use.

Mr. Cordeiro joined the Department of Regional Development and Environment as project chief of the OAS Plata Basin Program, in which Argentina, Bolivia, Brazil, Paraguay and Uruguay participated. In 1973 he was appointed Division Chief of the Department, a position that he has held since. He directs the technical cooperation activities of the Department for South America and, since 1984, directly supervises execution of the OAS Plurinational Project of Amazon Cooperation with the participation of Bolivia, Brazil, Colombia, Ecuador, Guyana, Peru, Suriname, and Venezuela.

He has served on several international commissions and advisory councils, and has also participated in various conferences and professional seminars throughout Europe and the Americas. He is the author of many technical publications and has been the principal contributor to numerous publications of the Department, especially those related to the Plata Basin and Amazon Basin cooperation projects.

Luis Veiga da Cunha

Dr. da Cunha has been Science Administrator in the NATO Scientific Affairs Division since 1983, currently having, among other duties, responsibility as Director of the Special Programme on The Science of Global Environmental Change. His scientific expertise is in the field of Environment and Natural Resources Planning and Management, with special emphasis on water resources problems. He is a civil engineer by training, having acquired his engineering degree and Ph.D. at the Technical University of Lisbon. He worked in Portugal as a professor and researcher until his appointment to NATO. During that period, he was involved in numerous research projects and consulting activities, lecturing in Portugal and various foreign countries.

He was a founding member and first President of the Portuguese Water Resources Association and Director of the International Water Resources Association. He has published 20 books and book chapters, and 70 technical and scientific papers. He was Portugal's Minister of Education in 1979.

Michael H. Glantz

Michael H. Glantz is a Senior Scientist and the Director of the Environmental and Societal Impacts Group, a program at the National Center for Atmospheric Research (NCAR). He is interested in how climate affects society and how society affects climate, especially how the interaction between climate anomalies and human activities affect quality of life issues. His research relates to African drought and desertification and food production problems and prospects; societal impacts of climate anomalies related to El Niño events, and the use of El Niño-related teleconnections to forecast these impacts; to developing methods of forecasting possible societal responses to the regional impacts of climate change; and the use of climate-related information for economic development. He also coordinated joint research in the Central Asian Republics of the former Soviet Union.

He received his B.S. in Metallurgical Engineering (1961) and an M.A. in Political Science (1963) from the University of Pennsylvania. After some years in industry (Westinghouse, Ford), he returned to the University of Pennsylvania and received a Ph.D. in Political Science in 1970. He has taught at the University of Pennsylvania, Lafayette College, and Swarthmore College. In 1974 he joined NCAR as a postdoctoral fellow, and in 1983 he was the first social scientist to became a Senior Scientist. He is a member of numerous national and international committees and advisory bodies related to environmental issues. In 1987 his *Scientific American* article on drought in Africa was given an award by World Hunger Year. In March 1990 he received the "Global 500" award from the United Nations Environment Programme.

He has edited several books and is the author of numerous articles on issues related to climate, environment, and policy. He has a biweekly newspaper column called "Environmental Minute." His most recent edited books are *Drought and Hunger in Africa*, (Cambridge University Press, 1987 and 1988), *Societal Responses to Regional Climate Change: Forecasting by Analogy* Westview Press, 1988), *ENSO Teleconnections Linking Worldwide Climate Anomalies: Scientific Basis and Societal Impact* (Cambridge University Press, 1990), and *Climate Variability, Climate Change and Fisheries* (Cambridge University Press, 1992).

Saleemul Huq

Dr. Saleemul Huq was born in Pakistan in 1952 and obtained his early schooling in Germany, Indonesia, and Kenya. He received a B.Sc. (Honors) in Botany from Imperial College at London University in the United Kingdom in 1975 and, later on, DIC and Ph.D. also in Botany from Imperial College, London University in 1978. He then taught in the Department of Botany, Dhaka University in Dhaka, Bangladesh, until 1984 when he became the founding Executive Director of the Bangladesh Centre for Advanced Studies, which is currently the biggest non-government scientific and policy research institute on environment in Bangladesh, with over 30 full-time professionals working on various projects.

Dr. Huq has also carried out research and teaching at Imperial College, UK, and was awarded the Monamara Fellowship from the World Bank, Washington DC, USA, in 1986 and the Duggan Fellowship from the Natural Resources Defense Council, USA, in 1989. He has been a guest lecturer at numerous international environmental conferences and has authored several books and numerous articles on national, regional, and international environmental issues. He has worked as an environmental consultant for a number of international organizations including UNDP, World Bank, Ford Foundation, DANIDA, NORAD and SIDA. In June 1992 he co-organized and ran an international workshop on Environment and Poverty in Rio de Janeiro, Brazil, during UNCED, which culminated in the creation of the Global Forum on Environment and Poverty with over 200 member organizations from all over the world. He is also the Coordinator of the Climate Action Network—South Asia (CANSA), which coordinates research and publications on climate-related issues in the South Asia region.

Robert S. Kandel

Dr. Kandel is the Director of Research (Senior Research Scientist) at the Centre National de la Recherche Scientifique (CNRS) of France. He received a B.A. from Harvard College in 1958 and a Doctorat d'Etat in Astrophysics from the Faculté des Sciences in Paris in 1967.

He has held teaching and research positions in astronomy and astrophysics at the Observatoire de Paris/Meudon (France), 1961-67; NAS/NRC Postdoctoral Research Associate, NASA/Goddard Institute for Space Studies (USA), 1967-69; Department of Physics and Astronomy, Boston University, 1969-1974; and Observatoire de Paris/Meudon, 1974-77.

Dr. Kandel's involvement in problems of climate and climate change and observation of Earth from space is represented by teaching and research positions at Service d'Aéronomie of CNRS, 1978-85; Laboratoire de Météorologie Dynamique of CNRS, Ecole Polytechnique, Palaiseau, since 1985. He has been Director of Research with the CNRS since 1985.

His areas of special interest are satellite observation of the Earth's radiation budget and clouds, in relation to the physics of climate; member of the NASA ERBE (Earth Radiation Budget Experiment) and CERES (Clouds and the Earth's

Radiant Energy System) science teams; of the European ICE (International Cirrus Experiment) and EUCREX (European Cloud-Radiation Experiment) science teams; and Principal Scientist for the French-Russian-German ScaRaB (Scanner for Radiation Budget) experiment.

Dr. Kandel is a member of the International Astronomical Union, American Geophysical Union, American Meteorological Society, European Geophysical Society, Comité National Français de Géodésie et Géophysique, International Radiation Committee, COSPAR (Committee for Space Research), Committee for Earth Observation, Mouvement Universel de la Responsabilité Scientifique; Ocean-Atmosphere section, Comité National de la Recherche Scientifique (1987-91); Earth-Atmosphere-Ocean working group, Centre National d'Etudes Spatiales (1988-92).

He is the author of numerous research papers, reports, and articles, and of two books: *Earth and Cosmos* (Pergamon, 1980); and *Le Devenir des Climats* (Hachette, Paris, 1990), winner of the 1990 Roberval Prize for a popular science book in French; in English, *Our Changing Climate* (McGraw-Hill, New York, 1992).

Some of his recent general-interest publications are: "Satellite observations of the Earth radiation budget and clouds" *Space Science Reviews,* **52**, 1-32 (1990); and "Changement climatique: Le débat se réchauffe" *La Terre Outragée - Les experts sont formels!* (eds. J. Theys and B. Kalaora, Editions Autrement, Paris), 104-110.

Gunnar E.B. Kullenberg

Dr. Kullenberg received a B.Sc. in mathematics and physics in 1963, a Ph.D. in physical oceanography in 1967, and a D.Sc. in oceanography in 1974 at Copenhagen. He has been a professor of oceanography since 1979 at the University of Copenhagen. From 1977-1979, he was a professor of oceanography at the University of Goteborg. He has been Senior Assistant Secretary of the IOC since 1985 and a member of the Royal Danish Academy of Sciences and Letters since 1980.

He has been involved in developing, organizing, and teaching graduate and undergraduate courses and has assisted in the planning of national and international scientific programs. He has written five books and has approximately 80 published articles. His areas of scientific interest are ocean currents, mixing and turbulence; optics; water exchange; modeling; coupling physical-biological processes; pollution, degradation and management problems; air-sea interaction and remote sensing.

From 1985 through 1988 he was Senior Assistant Secretary for the IOC, responsible for the marine pollution research and monitoring program; since 1986 he has worked on an interim basis for the IOC's ocean sciences program. He has been UNESCO Technical Secretary for GESAMP and liaison with the Division of Marine Sciences. He has developed and presented a course module at the World Maritime University. Since January 1989, he has been Secretary of the IOC.

Intergovernmental organizations in which Dr. Kullenberg has participated are the International Council for the Exploration of the Sea, Chairman of Working

Group on the Baltic; Advisory Committee on Marine Pollution, 1974-83, Chairman 1978-81; Consultative Committee 1978-81; Chairman 1983-85; Convener of ICES symposia 1972, 1981, 1984; organizer BOSEX 1977. He has been involved in the following United Nations bodies: Member, GESAMP 1973-82, Chairman 1975-78, Chairman two Working Groups, member of four others; participation in IAEA Review Committees; Chairman IAEA Advisory Group; participation in workshops organized by IOC, FAO, UNEP; participation in IOC meetings and GIPME sessions, Helsinki Convention, participation meetings and expert groups. He has also been involved in non-governmental organizations: SCOR, Danish national member, chairman Danish National Committee.

Dr. Kullenberg has helped to plan, implement, and analyze many large international programs, including the International Southern Ocean Studies, Joint Air-Sea Interaction Expedition, Coastal Upwelling Ecosystems Africa/Peru, Fladen Ground Expedition 1976, Baltic 1977, Nordic Expeditions to Arctic, Greenland, Mediterranean, Sargasso Sea, among others.

Jean Labrousse

Jean Labrousse attended secondary school in Toulouse and university in Paris. He holds the degree of Graduate Ingenieur de la Meteorologie. He began working with the Meteorologie Nationale in 1952 and became the data processing manager there in 1964.

In 1974 he joined the European Centre for Medium Range Weather Forecast (ECMWF) in Reading, United Kingdom, as Head of the Operation Department. In 1980 he became the Director of the ECMWF.

In 1982 he was appointed Director of Meteorologie Nationale. In 1986 he became Conseil Général des Ponts et Chaussées. In 1987 he became the Director of Research and Development at the World Meteorological Organization (WMO) in Geneva, Switzerland.

Since 1991, he has been Head of the Earth, Ocean, Space, and Environment Department of the French Ministère de la Recherche et de l'Espace.

Jag S. Maini

Dr. Maini received his Ph.D. degree in Ecology from the University of Saskatchewan in 1960. As a Research Associate with the Institute for Northern Studies of the same university, he led ecological survey expeditions in the sub-arctic region of central Canada during 1960 and 1961.

He joined the Canadian Forestry Service in 1962 as a Research Scientist in the Ontario Region at Richmond Hill. Dr. Maini moved to Ottawa in 1968 as Coordinator of Programs in Tree Biology and the Environment. From 1973 to 1978, he served as Regional Director, Laurentian Forest Research Centre in Quebec City, and returned to Ottawa in 1978 as Project Director with the Ministry of State for Science and

Technology. He has authored two books and more than ninety scientific and policy papers.

Dr. Maini was appointed as Director General, Policy, Corporate Planning Group of Environment Canada in 1979 and was on Executive Interchange with the International Institute for Applied Systems Analysis in Vienna from 1983 to 1985. He joined the Temporary Assignments Program of the Treasury Board in 1986 and was assigned as Special Advisor to the Associate Deputy Minister, Canadian Forestry Service. In 1987, Dr. Maini was appointed as Assistant Deputy Minister, Policy, and in 1990 as Assistant Deputy Minister, Forest Environment with Forestry Canada.

Jag Maini has frequently represented Canada nationally and internationally and served in many capacities on numerous national and international government, nongovernment and professional organizations. At present, he is serving on:

- Advisory Board on Global Change, The Royal Society of Canada.
- NATO's Special Programme Panel on Global Climate Change.
- Steering Committee, European Forum on Forest Protection, sponsored by the InterAction Council (composed of former Heads of State; Chairman: Helmut Schmidt, former Chancellor, FRG).
- Chairperson, Canadian Committee for the UNESCO Programme on Man and the Biosphere (MAB).
- Member, Natural Sciences Committee, Canadian Commission for UNESCO.
- Fellow and Member, Board of Directors, The Royal Canadian Geographical Society.
- Adjunct Professor, Faculty of Forestry, University of Toronto.
- Member, Advisory Board, The Centre for Plant Biotechnology, University of Toronto.
- Member, Advisory Board, Plant Biotechnology Institute, National Research Council.
- Advisory Committee of the Global Warming Science and Policy International Conference, SUPCON International, Woodbridge, Illinois.
- Member, Advisory Committee on Biosphere Reserves, UNESCO, Paris, France.
- Chief Negotiator, Forests, Canadian Delegation to the Earth Summit, Rio de Janeiro, Brazil, June 1992.

Edward L. Miles

Edward Miles received his Bachelor of Arts degree (Magna cum laude, Honors in History) from Howard University in June 1962 and was at the same time elected to Phi Beta Kappa. He received his Ph.D. in International Relations in August 1965 from the Graduate School of International Studies, University of Denver. Between September 1965 and June 1974, he was respectively Instructor, Assistant Professor, and Associate Professor of International Relations at the Graduate School of International Studies, University of Denver. In July 1974 he moved to the Institute of Marine Studies and the Graduate School of Public Affairs, University of Washington, as Professor of Marine Studies and Public Affairs. In July 1982 he became Director of the Institute (which became the School of Marine Affairs in September 1991) and will serve in that position until July 1993.

His areas of specialization are International Law and Organization; Science, Technology and International Relations; and Marine Policy and Ocean Management. He has been *inter alia* a Ford Foundation Fellow (1968-70), an International Affairs Fellow of the Council on Foreign Relations (1972-73), a James P. Warburg Fellow of the University Consortium for World Order Studies, Center for International Affairs, Harvard University (1973-74) and simultaneously a Senior Fellow at the Woods Hole Oceanographic Institution.

He has served as chairman of the Ocean Policy Committee, National Academy of Sciences/National Research Council (1974-79); member of the Executive Board, Law of the Sea Institute, 1972-81 and 1985-89 and President 1989-93; Chairman of the Legal and Institutional Task Group on the Implications of Disposal of High-Level Radioactive Waste into the Seabed and Advisor to the Executive Committee, Seabed Working Group, Nuclear Energy Agency, OCED, 1981-1987; and Chairman of the Advisory Committee on International Programs of the National Science Foundation, 1990-92.

He has also served as consultant to the United Nations, the Intergovernmental Oceanographic Commission of UNESCO, the Department of Fisheries of FAO, and the South Pacific Forum Fisheries Agency. He is the author of many studies on international organizations, international science and technology policy, and marine policy and ocean management.

Renat Perelet

Dr. Perelet's major subject of interest is international environmental management. He worked with the USSR Commission for UNEP from 1978 to 1985, participating in UNEP's governing council, including sessions as a member, deputy head of Soviet government delegations, and working closely with the Brundtland Commission.

He has taken part in many international meetings and panels in the environmental area. As a consultant, he has worked for UNEP in a review of Earthwatch, environment in the North-South dialogue, and in development of an environmental security concept. He has been involved in UNESCO's MAB Programme on natural resource use and human investment, UNECE economic activity and conservation of flora, fauna, and their habitats, IIASA, and USSR/Russian Parliament.

He has published about 100 papers and several books on international environmental management in both Russian and English.

Other accomplishments of Dr. Perelet include the following:

- Research Leader, Institute for Systems Analysis, Russian Academy of Science
- Associate Professor, Moscow State University
- Director, Center for Ecological Economics, Russian Committee for UNESCO's MAB Programme
- Member, Russian National Committee of the International Geosphere-Biosphere Programme
- Member, international standing committee of the Human Dimensions of Global Environmental Change Programme.

Martin F. Price

Martin Price received B.Sc. and M.Sc. degrees in environmental science from the Universities of Sheffield (England) and Calgary (Canada), and a Ph.D. in geography from the University of Colorado (USA). A major focus of his research has been the sustainable use of mountain resources in the Rocky Mountains and the Alps. He has worked in many interdisciplinary situations, including periods as an environmental and a museum consultant, and as a technical writer.

From 1988 to 1991, as a postdoctoral fellow with NCAR's Environmental and Societal Impacts Group, he worked on the human dimensions of global environmental change (HDGEC). For much of this period, he was Secretary of the Standing Committee on HDGEC of the International Social Science Council, which produced a framework for global social science research on HDGEC.

In 1991, he became Scientific Director of the International Centre for Alpine Environment (France), a regional organization for the coordination of scientific research and communication in the European mountains. Subsequently, he conducted a review of research activities in the mountains of eastern Europe and the former USSR within the scope of UNESCO's Man and the Biosphere (MAB) programme. He is now working at the Environmental Change Unit, University of Oxford (UK) on pilot studies for the inventory of sources and sinks of greenhouse gases in North and West Africa.

Rui Roda

Rui Roda graduated in finances from the Economics and Management Institute, Lisbon, Portugal.

From 1971 to 1975 he worked as financial analyst at the Instituto de Credito de Angola (Angola Development Bank).

From 1975 to 1981 he was financial analyst at the Banco de Nacional in Lisbon, a state-owned investment bank.

Between 1981 and 1987 he was financial manager at IFADAP, a public institute for agriculture and fisheries financing.

He was consultant and advisor in the cabinet of the Secretary of State for the Treasury in 1979, the Secretary of State for Finances in 1980-1985, and Secretary of State for the Environment in 1985-1987.

Currently he is a member of the Board of Directors of EPAL, the Portuguese Water Resources Company, responsible for the establishment of regional water companies in conjunction with legal authority.

Recently he was named Director-General of Water Resources at the Portuguese National Water Authority.

Naresh C. Singh

Dr. Singh received a B.Sc. in Natural Sciences from University of Guyana, M.Sc. in Soil Science and Agricultural Chemistry from I.A.R.I., New Delhi, and Ph.D. in Environmental Chemistry from University of the West Indies, Jamaica.

He is presently Director of the Caribbean Environmental Health Institute, a specialist technical institution of the Caribbean Community in environmental management research and monitoring.

Among Dr. Singh's accomplishments are:

- Head/Alternate Head, St. Lucia Delegation to INC - Climate Change
- Adjunct Professor - Department of Civil Engineering, University of Waterloo
- Visiting Lecturer - University of the West Indies, Kingston, Jamaica
- Member - Centre for Sustainable Infrastructure, McMaster University, Ontario, Canada
- Member - Commonwealth Group of Experts on the Environment

He has over fifty publications in areas of environmental science, policy, economics and management.

He is a member of the following professional organizations:

- Royal Society of Chemistry (MRSC)
- IUCN Commission of Environmental Strategy and Planning
- International Society of Ecological Economics.

Ödön Starosolszky

Dr. Starosolszky graduated in hydraulic engineering from the Technical University of Budapest (Hungary) in 1954. He has, since that time, been engaged mostly in research, but also in education and technical administration (at the National Water Authority). He was appointed General Director (1991) of Hungary's Water Resources Research Centre (VITUKI). His research work has covered the areas of hydrology (forecasting, storage, urban hydrology), hydraulics (e.g., ice), hydrotechnical engineering (e.g., flood control), hydrometry (discharge measurements, standardization). His broad-ranging professional activities are reflected in his publication of more than 200 papers. His major publications include the Civil Engineering Hydraulics (1970), Ice in Hydraulics Engineering (1971), and Applied Surface Hydrology (1988).

He is the chairman of the Committee on Water Sciences of the Hungarian Academy of Sciences and of the Committee on International Relations of the Hungarian Hydrological Society. He has been lecturing in hydrometry and hydraulics at the Budapest International Postgraduate Course. Since 1960, he has been active in scientific international organizations and associations, such as UNESCO, WMO, IAHS, ICID. He was vice president of IAHR. He is the hydrological advisor to the Permanent Representative of Hungary with WMO and has recently finished his term as president of WMO Commission for Hydrology (1984–92). He has been a UNDP and WMO consultant on hydrologic and hydraulic measurements in developing countries (e.g., Sri Lanka, Nigeria, India, Egypt).

He is a team leader for the research program on the Impact of Climate Change on Hydrological Parameters, which is supported by the Hungarian Science Foundation (1991-94).

Sam Tewungwa

Mr. Tewungwa is a graduate of the Imperial College of Science and Technology (London University) and the University of Nairobi (Kenya). Mr. Tewungwa is currently a Senior Programme Officer in the World Meteorological Organization (WMO)/United Nations Environmental Programme (UNEP) Intergovernmental Panel on Climate Change (IPCC), where he handles most of the impacts and scientific aspects of the Panel. He was seconded to the Panel by UNEP in late 1988.

As Director-General of the former East African Meteorological Department of the former East African Community (1968-1977), and Permanent Representative of Kenya, Tanzania, and Uganda with the WMO, he was responsible for the development and administration of meteorological services in the East African region. His main responsibility was to advise the three governments on the meteorological aspects of agriculture, water resources management, air and marine transport.

In his capacity as Permanent Representative of the three East African States with WMO, Mr. Tewungwa was elected to the Executive Committee (now Council) of WMO in 1971 and re-elected in 1975.

During 1977-1980, he was a Scientific Officer in the WMO, responsible for the development and implementation of the organization's programmes related to desertification and renewable sources of energy.

As a Minister of Regional Cooperation in the Republic of Uganda during 1980-1985, among other duties, Mr. Tewungwa was responsible for:

- Negotiations of a Treaty for the establishment of the Preferential Trade Area (PTA) for Eastern and Southern Africa;
- The management of the Kagera River Basin Organization, a regional organization involving Uganda, Tanzania, Burundi, and Rwanda;
- Negotiations for the division of the assets and liabilities of the former East African Community between the former members;
- Negotiations for Uganda's membership in the Undugu Club, a regional political organization involving Uganda, Egypt, Sudan, Rwanda, Burundi and Zaire.

Mr. Tewungwa has represented Uganda, UNEP and WMO at many international conferences.

Jean-Paul Troadec

As a fishery biologist at France's ORSTOM, Jean-Paul Troadec worked in the 1960s on a survey and assessment of Gulf of Guinea fishery resources.

In 1971 he joined the Department of Fisheries in the FAO Headquarters in Rome, where he served as technical secretary of working parties on fisheries statistics, stock assessments, and management of FAO regional fisheries bodies (GFCM, CECAF, etc.) and provided technical backing in the same fields to fisheries development projects executed by FAO in West Africa and the Mediterranean.

During that period, he became interested in understanding fisheries development conditions in tropical areas, as well as in the change in fisheries management requirements and approaches following the full exploitation of world fishery resources and the emergence of the New Ocean regime. From 1978 to 1982, he was head of a Fisheries Development Planning Service in the FAO Fisheries Department.

In 1982 he became Director of the French Scientific and Technological Institute for Marine Fisheries and, in 1984, Deputy Director General and Director of Living Resources in the French Institute of Research for the Exploitation of the Sea.

In 1989, he was responsible for the preparation of the Study of International Fishery Research, a project jointly financed by the World Bank, UNDP, FAO, the European Community Commission and a dozen bilateral development agencies, to analyze the causes of the high failure rate experienced by development projects implemented in the developing world in the field of fisheries and aquaculture, to determine the kinds of research needed to contribute to upgrading project performances, and to propose approaches for developing necessary investigations.

From 1991 to 1992, he assisted the Government of Morocco in the preparation of fisheries management plans and in the assessment and design of institutions that are needed to improve fisheries efficiency and fishery management effectiveness.

His interests in fisheries research cover the historical development of production systems—including the use and conservation of marine environments, the related changes in potential inputs of scientific disciplines, and the institutional innovations needed to improve fisheries performances following the global emergency of resource scarcity.

He presented a Ph.D. thesis on the biology and dynamics of West African croakers at the University of Marseilles and edited a book on fisheries management *L'homme et les ressources halieutiques,* (IFREMER, 1989).

Ilter Turan

Dr. Turan received his B.A. from Oberlin College (Ohio) in 1962 and his M.A. in Government from Columbia University (New York) in 1964. He received his Doctorat des Sciences Economiques from the Faculty of Economics of Istanbul University in 1966.

He has been Assistant Professor of Political Science (1966-70), Associate Professor of Political Science (1970-76), and Professor of Political Science (1976-84) in the Faculty of Economics of Istanbul University. Since 1984, he has been a member of the Faculty of Political Science at Istanbul University where he directs the Center for the Study of the Middle East and the Balkans, and heads the recently established Department of International Relations.

He has held visiting appointments at various American institutions including the Universities of Iowa, Wisconsin, Arizona, Kentucky, and California (Berkeley).

He is the author of books and articles in English and Turkish on Comparative Politics, Political Culture and Political Behavior, Policy Making, Foreign Policy, and International Relations.

Arild Underdal

Arild Underdal is professor of political science at the University of Oslo, Norway, and director of research at the Center for International Climate and Energy Research (CICERO). His main fields of research are international negotiations, international resource and environmental management, and foreign policy decisionmaking. His books include *The Politics of International Fisheries Management: The Case of the Northeast Atlantic* (1980); *Oljevirksomheten og fiskerinaeringens interesser* (with S. Andresen, 1983); *Overordnet styringsinstans og sideordnet part* (with R.J. Sørensen, 1992); and *Rationality and Institutions* (co-editor, with R. Malnes, 1992). He chairs a new study group on the politics of global environmental change, under the International Political Science Association, and also one of the task forces under the European Science Foundation *Environment, Science and Society* program.

The Center for International Climate and Energy Research (CICERO) was established by the Norwegian Government to form part of a worldwide research network in support of climate and energy-related issues. The orientation of CICERO is toward the policy side of climate change, and its fields of research include (a) alternative global regimes, cost-effectiveness and economic consequences; (b) the role of international organizations and institutions; (c) conflicts and compromises in international climate policy; (d) the link between climate change and development; and (e) international law and different global regimes.

James L. Wescoat, Jr.

Dr. Wescoat is an assistant Professor of Geography at the University of Colorado at Boulder, USA. He is a member of the Environment and Behavior Program in the Institute of Behavioral Sciences, and on the Advisory Board of the Natural Hazards Center at that University. He has received fellowships and awards from the National Science Foundation, the Rockefeller Foundation for the Humanities, and Dumbarton Oaks.

Dr. Wescoat earned his doctoral degree in geography from the University of Chicago with a dissertation on water conflict and cooperation in the Colorado River basin, titled *Integrated Water Development* (1984). He has published related articles on the relationship between salinity control and water rights patterns, and on long-term change in irrigation systems in North America.

He has recently completed a three-year study of the potential impacts of climate change in the Indus River basin of Pakistan with the Pakistan Water and Power Development Authority. That case study was part of a comparative investigation of five complex river basins which included the Mekong, Nile, Uruguay, and Zambezi rivers. In addition to a national modeling report for the Indus, that study has yielded an article on four conceptual approaches to climate impact assessment in the Indus basin. Dr. Wescoat also published a recent analysis of the draft United Nations articles on the law of non-navigable uses of international watercourses. Current research is on long-term water development in South Asia and the American West, with special emphasis on the geographical interactions between those regions which have shaped current water management problems.

Citation: A/CONF.151/PC/100/Add.26
Item 2(c) of Plenary Session
Item 4 of provisional agenda of Working Group III

ROLE OF REGIONAL ORGANIZATIONS AND CO-OPERATION
(Section IV, Chapter 7 of Agenda 21)

INTRODUCTION

1. This document responds to decision 3/2 of the Preparatory Committee in addressing the role of regional organizations and activities in relation to capacity building. Although all chapters of Agenda 21 include references to regional co-operation and many programme areas also suggest such actions, it was deemed appropriate to prepare a brief paper on the role of regional organizations and co-operation. Other documents of relevance to this issue are: "National Mechanisms and International Co-operation for Capacity Building in Developing Countries" as well as the synthesis report on national level activities as reflected in the national reports, which often refer to regional or inter-country issues.

PROGRAMME AREA

2. While most of the follow up actions under Agenda 21 will be dealt with at the national level, almost all programme documents refer to regional requirements and co-operation, the need for regional centres of excellence, exchange of information, data collection, etc. National level capacity of countries is vital to the implementation of national policies, plans and programmes, and these capacities need to be strengthened. Regional activities are expected to increase because of the nature of environmental problems that know no boundaries and thus regional organizations as well should be strengthened or restructured.

Objectives

Long-term objectives:

3. Consolidation and harmonization of regional and sub-regional activities and actions. The large number of follow-up actions require better co-ordination than ever before. The core organizations within the United Nations system are UNDP, UNEP and the Economic Commissions. In addition, specialized agencies as well as regional development banks have an important role to play. Modification and strengthening of mandates of these agencies/organizations may result.

Short term objectives:

a. Each Member State to complete its own Agenda 21, for implementation of decisions taken at UNCED, by 1994. It is equally important for regional organizations, within the same time frame, to review and prepare their agencies in co-ordination with national priorities;

b. The Secretary-General of the United Nations prepare a complementary report to be submitted to the fiftieth session of the General Assembly on consultations among regional organizations, international organizations, donor agencies and private sector partners containing proposals for regional level co-ordination, with clearly defined responsibilities and possible changes to mandates.

Activities

Review and strengthen regional strategies for sustainable development

4. Organizations functioning at the regional level and regional programmes should address the issue of how to support, improve and strengthen regional and sub-regional consultative processes. The purpose is to assist with regional and sub-regional programme implementation in support of governments which are preparing policies and programmes needed for the next generation to implement its Agenda 21, at all levels in society. The Economic (and Social) Commissions and regional programmes of the UN system, as well as other regional organizations, should review the need for modification of ongoing activities, as appropriate in light of the Agenda 21, once approved.

Strengthen existing regional organizations

5. There are a variety of regional organizations ranging from political defense organizations and economic market organizations, to regional environmental entities. While these all have their mandate, most will require some internal adjustments to meet the new challenges introduced in Agenda 21. Therefore the following is proposed:

a. Organizations with an economic orientation should strengthen the environmental and sustainable development aspects of their activities and programmes;

b. Organizations with an environmental orientation should strengthen the developmental aspects of their activities;

c. The capacities of the relevant regional organizations and institutions should be strengthened to enable them to carry out these new and expanded functions as part of the capacity building programme referred to in A/CONF.151/PC.100/Add 11.

Harmonization and co-operation between regional organizations

6. Efficient co-operation between regional organizations and programmes of the United Nations system, as well as other regional organizations dealing with sustainable development, is necessary to avoid duplication of efforts, as well as for reasons of economy. Bilateral aid agencies should be encouraged to co-operate more closely with these regional entities.

7. The Economic Commissions of the United Nations system, UNEP and UNDP form the core group, and together with the regional development banks, should bear special responsibility for such co-ordination.[1]

8. The Economic Commissions of the United Nations system with mandates essentially for economic and social development have a particular responsibility in these fields which should be complemented with UNEP's know-how and role in the field of environment and UNDP's designated role as co-ordinator of United Nations system technical co-operation. The Economic Commissions have taken steps to modify their mandates in relation to environmental matters. These United Nations system bodies all report to the Economic and Social Council (ECOSOC) of the United Nations where the respective areas of responsibility could be reviewed during 1994 prior to the Secretary

[1] A separate document prepared by the Economic Commissions will be available at the Fourth Preparatory Committee meeting.

General's report to the fiftieth General Assembly. The regional development banks, whilst having close contacts with the United Nations system and having a different reporting responsibility, could provide financial support and know-how. Their role in this undertaking would have to be clarified.

9. The Member States belonging to the regional entities concerned should agree on the appropriate co-ordinating mechanism at the regional level.

Improve the process of joint action at regional level.

10. Co-ordination should be improved between these organizations and their respective activities. What is needed is more 'joint action' to address effectively sustainable development. This should also include the regional development bank, the World Banks and bilateral aid agencies in an appropriate way, perhaps in the form of round-table or special regional or sub-regional consultative meetings.

11. UNDP should be designated as the lead agency to co-ordinate global action of the core group, given its ongoing capacity building initiative and national level responsibility as well as involvement, for instance, in the African Capacity Building Initiative. Within each region the economic commissions should be designated as support agents to gather the information required for the report.

12. Alternatively a lead agency from outside the UN organization could be chosen, the main concern being that the organization have the capacity and desire to undertake the study. The support agencies could also be non-UN agencies and may vary from region to region.

Co-ordinate and harmonize sub-regional activities

13. Particular attention should be given to sub-regional activities and programmes. These activities, often based on eco-geographical realities, emerging naturally from within the communities, form strong and viable building blocks. Such activities and programmes are often linked to river basins, island countries, desert countries, etc. They are often strong enough to survive even armed conflicts. The review of regional concerns and capacities should not be restricted by existing political and geographical boundaries. Whilst recognizing the sovereignty of nations, environmental concerns within countries and their neighbors should allow for the formation of new groupings for the purpose of addressing sustainable development.

14. The role of the non-government sector on a regional level should also be included in an effective manner.

Means of Implementation

15. For the purpose of this study it is suggested that the lead agency and the support agencies, if they are part of the UN system, be provided with funding for the hiring of temporary staff or consultants.

List of Participants

NATO Advanced Research Workshop

The Role of Regional Organizations in the Context of Climate Change

5–8 October 1992
Paris, France

Michael H. Glantz [Organizer]
Environmental & Societal Impacts Group
National Center for Atmospheric Research
PO Box 3000
Boulder, CO 80307, USA

James P. Bruce
1875 Juno Avenue
Ottawa, Ontario
Canada K1H 6S6

Lynton K. Caldwell
Dept. of Political Science
University of Indiana
Bloomington, IN 47401, USA

Newton V. Cordeiro
Organization of American States (OAS)
17th St. & Constitution Ave. NW
Washington, DC 20006, USA

L.V. da Cunha
Programme Director
Scientific & Environmental Affairs Division, NATO
B-1110 Brussels, Belgium

Saleemul Huq
Executive Director
Bangladesh Centre for Advanced Studies
620 Road 10A (New)
Dhanmondi, GPO Box 3971
Dhaka 1205, Bangladesh

Robert S. Kandel
Laboratoire de Meteorologie Dynamique du CNRS
Ecole Polytechnique
Route Departmentale 36
91128 Palaiseau Cedex, France

Gunnar E.B. Kullenberg
Secretary, Intergovernmental Oceanographic Commission
UNESCO
7, place de Fontenoy
75700 Paris, France

Jean Labrousse
Departement Terre, Ocean, Espace et Environnement
1, rue Descartes
75231 Paris, France

Jag S. Maini
Asst. Deputy Minister
Forest Environment
Forestry Canada
Ottawa, Canada K1A 1G5

Edward Miles
Director, School of Marine Affairs, HF-05
3707 Brooklyn Ave. NE
University of Washington
Seattle, WA 98195, USA

Renat Perelet
Institute for Systems Analysis
Russian Academy of Sciences
Prospect 60-Let Oktyabrya, 9
112317 Moscow, Russia

Natalie Philippon
Intergovernmental Oceanographic Commission
UNESCO
7, place de Fontenoy
75700 Paris, France

Martin F. Price
(Rapporteur)
University of Oxford
Environmental Change Unit
1a Mansfield Rd.
Oxford OX1 3TB UK

Rui J.M.R. Roda
Empresa Portuguesa das Aguas Livres (EPAL)
Av. da Liberdade, 24
1122 Lisbon, Portugal

Naresh C. Singh
Interenational Institute for Sustainable Development
161 Portage East
Winnipeg, Manitoba
Canada R3B 0Y4

Ödön Starosolszky
VITUKI
Kassay Jenö u. 1
H-1095 Budapest, Hungary

Sam Tewungwa
IPCC Secretariat
World Meteorological Organization
C.P. No. 2300
1211 Geneva 2
Switzerland

Jean-Paul Troadec
Centre ORSTOM
BP 70
29280 Plouzane Cedex
France

Ilter Turan
Chair, International Relations
Faculty of Political Science
Istanbul University
Istanbul, Turkey

Arild Underdal
Center for International Climate & Energy Research (CICERO)
PO Box 1066
Blindern, N-0316 Oslo, Norway

James L. Wescoat, Jr.
Department of Geography
University of Colorado
Boulder, CO 80309, USA

Secretariat:
D. Jan Stewart
Isabel Dolan
Sylvie Vacheret

Suggested Bibliography

Agenda 21, 1992: Role of regional organizations and cooperation. UNCED Proceedings, Item 2(c) of Plenary Session, Item 4 of provisional agenda of Working Group III. (Section IV, Chapter 7 of Agenda 21).

Antal, E., and O. Starosolszky, 1990: *Role of the Climate and Climate Change in the Life of Hungary*. Prepared on behalf of the Hungarian Ministry for Environment and Regional Policy, Budapest, Hungary.

Bjorkbom, L. and U. Svedin, 1992: Environmental security and the long-term European situation. In *Yearbook 1990-91: Towards a New European Security Order*, Eds. B. Huldt and G. Herolf. The Swedish Institute of International Affairs.

Bruce, J.P., 1992: *Meteorology and Hydrology for Sustainable Development*. WMO No. 769, Secretariat of the WMO, Geneva, Switzerland.

Bruce, J.P., 1992: Challenges of the decade: Natural disasters and global change. Opening address for symposium, "The World At Risk: Natural Hazards and Climate Change," 14-16 January 1992, at MIT, Cambridge, Massachusetts, USA. Appearing in forthcoming Conference Proceedings volume.

Caldwell, L.K., 1993: The Challenge of appropriate institutions with particular reference to the Great Lakes Governance. In *Proceedings of the Canada-United States Inter-University Seminar on Great Lakes Governance*, Ed. M.J. Donahue. Buffalo, NY: State University of New York, 11-15.

Caldwell, L.K., 1992: *Between Two Worlds: Science, the Environmental Movement and Policy Choice*. (paperback) Cambridge, U.K.: Cambridge University Press.

Caldwell, L.K., 1992: Transboundary conflicts: Resources and environment. In *The Canada-United States Relationship: The Politics of Energy and Environmental Coordination*, Ed. J. Temco. Westport, CT: Prosper, 15--40.

Caldwell, L.K., 1990: Regional arrangements: Bilateral and multilateral agreements. In *International Environmental Policy: Emergence and Dimensions*, L.K. Caldwell. London, U.K.: Duke University Press, 129-167.

Caldwell, L.K., Ed., 1988: *Perspectives on Ecosystem Management for the Great Lakes*. Albany, NY: State University of New York Press.

Caldwell, L.K., 1985: Binational responsibilities in a shared environment. In *Canada and the United States: Enduring Friendship, Persistent Stress*, Eds. C.F. Doran and J.H. Sigler. Englewood Cliffs, NJ: American Assembly and Council on Foreign Relations (Prentice-Hall), 203-230.

Döös, Bo R., 1991: *Environmental Issues Requiring International Action.* IIASA, RR-91-16, October. Laxenburg, Austria: IIASA (International Institute for Applied Systems Analysis).

Dublin Statement of Water and Sustainable Development. Statement adopted at closing session of International Conference on Water and the Environment, 26-31 January 1992, Dublin, Ireland.

Enquete Commission "Protecting the Earth's Atmosphere" of the 12th German Bundestag (Ed.), 1992: *Climate Change -- A Threat to Global Development: Acting Now to Safeguard the Future.* Bonn, Germany: Economica Verlag; Karlsruhe: Müller.

Enquete Commission of the 11th German Bundestag, 1991: *Protecting the Earth: A Status Report with Recommendations for a New Energy Policy,* Vols. 1 and 2. Bonn, Germany: Dt. Bundestag, Referat Öffentlichkeitsarbeit.

Enquete Commission of the 11th German Bundestag, 1990: *Protecting the Tropical Forests: A High-Priority International Task.* Bonn, Germany: Dt. Bundestag, Referat Öffentlichkeitsarbeit.

Enquete Commission of the 11th German Bundestag, 1989: *Protecthing the Earth's Atmosphere: An International Challenge: Interim Report of the Study Commission,* "Preventive Measures to Protect the Earth's Atmosphere." Bonn, Germany: German Bundestag, Publ. Sect.

Galtung, J., 1982: *Environment, Development, and Military Activity: Towards Alternative Security Doctrines.* Oslo, Norway: Universitetsforlaget.

Gebremedhin, N., S. Lodgaard, J. Opschoor, and R. Perelet, 1989: *Environmental Security.* Nairobi, Kenya: United Nations Environment Programme.

Glantz, M.H., A.Z. Rubinstein, and I. Zonn, 1992: Tragedy in the Aral Sea basin: Looking back to plan ahead. Paper to be presented at Villanova University seminar, "Central Asia: Its Strategic Importance and Future Prospects," 29-30 October 1992. In preparation.

Glantz, M.H., 1992: Global warming impacts on living marine resources: Anglo-Icelandic Cod Wars as an analogy. In M.H. Glantz (Ed.), *Climate Variability, Climate Change and Fisheries.* Cambridge, UK: Cambridge University Press, 261-290.

Glantz, M.H. (Conference Organizer), 1990: *Report of the Workshop "On Assessing Winners and Losers in the Context of Global Warming,"* held 18-21 June 1990 in St. Julians, Malta, sponsored by UNEP and the National Center for Atmospheric Research, Boulder, Colorado, USA.

Homer-Dixon, T.F., 1991: On the threshold: Environmental changes as causes of acute conflict. *International Security,* **16**(2), 76-116.

Houghton, J.T., B.A. Callander and S.K. Varney, 1992: *Clcimate Change 1992: The Supplementary Report to the IPCC Scientific Assessment.* Report prepared for IPCC (Intergovernmental Panel on Climate Change) by Working Group I. Cambridge, U.K.: Cambridge University Press.

Houghton, J.T., G.J. Jenkins and J.J. Ephraums: 1990: *Climate Change: The IPCC Scientific Assessment.* Report prepared for IPCC by Working Group I. Cambridge, U.K.: Cambridge University Press.

Intergovernmental Panel on Climate Change, 1991: *Climate Change: The IPCC Response Strategies.* Report prepared for IPCC by Working Group III. Washington, DC: Island Press.

Jäger, J., and H.L. Ferguson (Eds.), 1991: Second World Climate Conference, Geneva, Switzerland, 7 November 1990. *Climate Change: Science, Impacts and Policy,* Cambridge, UK: Cambridge University Press, 497-503.

Kaczmarek, Z. (Chair), 1991: Task Group 1: Climate, hydrology and water resources. In J. Jäger and H.L. Ferguson (Eds.), *Climate Change: Science, Impacts and Policy,* Proceedings of the Second World Climate Conference, Cambridge, UK: Cambridge University Press, 435-438.

Kandel, R., 1990: *Our Changing Climate.* New York, NY: McGraw-Hill.

Keyfitz, N., 1991: *Population and Development within the Ecosphere: One View of the Literature.* IIASA, RR-91-14, August. Laxenburg, Austria: IIASA (International Institute for Applied Systems Analysis).

Maier-Regaud, G., 1991: *Ecological Economics and Global Change.* Bonn, Germany: Institute für Europeache Umwelpolitik, September.

Maini, J.S., 1991: Towards an international instrument of forests. Background paper prepared for Informal Intergovernmental Consultation, 21-22 February 1991, Geneva, Switzerland.

Maini, J.S., 1989: Forests and atmospheric change. In *The Changing Atmosphere: Implications for Global Security,* Conference Proceedings, 27-30 June 1988, Toronto, Canada. WMO No. 710, Secretariat of the WMO, Geneva, Switzerland, 193-208.

Maini, J.S. and O. Ullsten, 1991: Conservation and sustainable development of forests world-wide: Issues and opportunities. Discussion paper presented at an international workshop on "The Conservation and Utilization of World Forests," 17-20 October 1991, Woods, Hole, Massachusetts.

Miles, E.L., 1983: On the roles of international organizations in the new ocean regime. In Choon-ho Park (Ed.), *The Law of the Sea in the 1980s*. Honolulu, Hawaii: University of Hawaii Press, 383-445.

Miles, E.L., 1982: International arrangements to facilitate and coordinate future marine science programs. In *United States Interests and Needs in the Coordination of International Oceanographic Research,* Washington, DC: National Academy Press, 115-120.

Miles, E.L., 1978: On the utility of regional arrangements in the new ocean regime. In Douglas and Johnston (Eds.), *Regionalization of the Law of the Sea,* Boston, MA: Ballinger Publishing Co., 255-275.

Painchaud, P., 1992: *Geopolitical Perspectives on Environmental Security*. The Studies and Research Center on Environmental Policies (GERPE), Université Laval, Quebec, Canada. Cahier du GERPE, No. 92-05.

Perelet, R., 1991: *Managing Global Change: The Prospects for the 1992 UN Conference on Environment and Development.* Seminar report on Global Change and Finland, Research Report 7/90. Helsinki, Finland: The Finnish Institute of International Affairs.

Perelet, R., 1991: *Strengthening the Role of the UN in Furthering Global Security, Environmental Protection and Sustainable Development.* Doc. PA 33/9/12/USSR, A only. 33rd Plenary Assembly, World Federation of UN Associations, Barcelona, Spain, 18-23 November 1991.

Perelet, R. and V. Iakimetz, 1992: Environmental parity. (In Russian.) *Spasenie* (Ecological weekly), Vol. 1, No. 17, January.

Perelet, R., S. Evteev and V. Voronin, 1989: Environmental security of sustainable development. In *The Future Role of the United Nations in an Interdependent World,* Ed. John P. Renninger, 162-171. The Netherlands: Nartinus Nijhof Publishers.

Rogers, P., 1991: International river basins: Pervasive unidirectional externalities. Paper presented at a conference on "The Economics of Transnational Commons," Universita di Siena, Italy, 25-27 April 1991.

Singh, N.C., 1992: Small islands collaboration and the climate change negotiations. Paper prepared for conference, "Islands 2000 - The World of Islands - What Development to the Year 2000?", 19-24 May 1992, Naxos, Italy: Giardini.

Singh, N.C., 1992: Environmental problems and opportunities in small open economies. Paper prepared for symposium, "Small States: Problems and Opportunities in a World of Rapid Change," 25-27 March 1992, St. Kitts.

Solanes, M., 1992: Legal and institutional aspects of river basin development. *Water International*, **17** (3), 116-123.

Troadec, J.-P., 1989: Les pêches et les cultures marines face à la rareté de la ressource. In J.P. Troadec (sous la dir.) - *L'homme et les ressources halieutiques. Essai sur l'usage d'une -ressources commune renouvelable.* IFREMER, Paris, 21-66.

Troadec, J.-P., 1992: L'amenagement des pêches: Une necessite irrealisable? (In French.) Paper presented at Institut international des océans, Séminaire sur l'aménagement des ressources marines méditerranéennes, Agadir (Maroc), 16 novembre - 11 décembre 1992.

Turan, I., 1992: Turkey and the Middle East: Problems and solutions. Mimeo. Department of International Relations, Istanbul University, Istanbul, Turkey.

Underdal, A., 1990: Designing political feasible solutions. Chapter in R. Malnes and A. Underdal (Eds.), *Rationality and Institutions: Essays in Honour of Knut Midgaard,* Studies in Political and Government 1, Scandinavian University Press, 221-254.

Underdal, A., 1990: International cooperation and political engineering. In S.S. Nagel (Ed.), *Global Policy Studies: International Interaction toward Improving Public Policy,* MacMillan, 98-120.

Water International, 1990: *Special Issue on Water, Peace & Conflict Management*, **15** (4), December.

Wescoat, J.L., Jr., 1992: Beyond the river basin: The changing geography of international water problems and international watercourse law. *Colorado Journal of International Environmental Law and Policy,* **3**, 301-330.

Wescoat, J.L., Jr., 1991: Managing the Indus River basin in light of climate change: Four conceptual approaches. *Global Environmental Change,* **1**, 381-395.

Wescoat, J.L., Jr., 1990: Common law, common property, and common enemy: Notes on the political geography of water resources management for the Sundarbans area of Bangladesh. *Agriculture and Human Values,* **7**, 73-87.

GLOSSARY OF ACRONYMS

AAAS	American Association for the Advancement of Science
ACMAD	African Centre of Meteorological Application for Development
AGRHYMET	Agriculture, Hydrology and Meteorology
AOSIS	Association of Small Island States
ARW	Advanced Research Workshop
ASEAN	Association of South-East Asian Nations
CANSA	Climate Action Network - South Asia
CARIPOL	Regional Marine Pollution Program of IOCARIBE
CECAF	Fishery Committee, Eastern Central Atlantic
CEHI	Caribbean Environmental Health Institute
CERES	Clouds and the Earth's Radiant Energy System
CFCs	Chlorofluorocarbons
CICERO	Center for International Climate and Energy Research
CIDA	Canadian International Development Agency
CIECC	Inter-American Council for Education, Science and Culture
CIES	Inter-American Economic and Social Council
CILSS	Permanent Interstate Committee for Drought Control in Sahel
CIMH	Caribbean Institute of Meteorology and Hydrology
CINCWIO	Central & Western Indian Ocean
CMO	Caribbean Meteorological Organization
CNRS	Centre Nationale de la Recherche Scientifique
CO_2	Carbon Dioxide
COMECON	Council for Mutual Economic Assistance (also known as CMEA)
COSPAR	Committee for Space Research
CPPS	Comision Permanente del Pacifico Sur
CSCE	Conference on Security and Cooperation in Europe
DANIDA	Danish International Development Agency
DRDE	Department of Regional Development and Environment
DWFNs	Distant Water Fishing Nations
EC	European Community
ECAFE	Economic Commission for Asia and the Far East
ECMWF	European Centre for Medium Range Weather Forecast
ECOSOC	Economic and Social Council (UN)
EEC	European Economic Community
EEZ	Exclusive Economic Zones
ENSO	El Niño-Southern Oscillation
ERBE	Earth Radiation Budget Experiment (NASA)
ESCAP	Economic Social Commission for Asia and the Pacific
ESIG	Environmental and Societal Impacts Group
ETAP	Expanded Technical Assistance Program
EUCREX	European Cloud-Radiation Experiment
FAO	Food and Agriculture Organization of the UN
FCCC	Framework Convention on Climate Change

FINNIDA	Finnish International Development Agency
GATT	General Agreement on Tariffs and Trade
GCM	General Circulation Model
GDP	Gross Domestic Product
GEC	Global Environmental Change
GESAMP	Group of Experts on Scientific Aspects of Marine Pollution
GEWEX	Global Energy and Water Cycle Experiment
GFCM	General Fisheries Council, Mediterranean
GHGs	Greenhouse Gases
GIPME	Committee for Global Investigation, Pollution of Marine Environment
HDGEC	Human Dimensions of Global Environmental Change
IAHS	International Association of Hydrological Sciences
ICE	International Cirrus Experiment
ICES	International Council for the Exploration of the Sea
ICID	International Commission on Irrigation Drainage
ICSPRO	Intersecretariat Committee on Scientific Programmes Related to Oceanography
ICSU	International Council of Scientific Unions
IDB	Inter-American Development Bank
IDNDR	International Decade for Natural Disaster Reduction
IGBP	International Geosphere-Biosphere Program
IGO	Intergovernmental Organizations
IIASA	International Institute for Applied Systems Analysis
IISD	International Institute for Sustainable Development
IJC	International Joint Commission
IMF	International Monetary Fund
INC	Intergovernment Negotiating Committee
IOC	Intergovernmental Oceanographic Commission
IOCARIBE	IOC Sub-Commission for the Caribbean and Adjacent Regions
IOCEA	IOC Regional Committee for the Central Eastern Atlantic
IOCINCWIO	IOC Regional Committee for the Cooperative Investigations in the North and Central Western Indian Ocean
IOCINDIO	IOC Regional Committee for the Central Indian Ocean
IPCC	Intergovernmental Panel on Climate Change (UNEP/WMO)
IWC	International Whaling Commission
JICA	Japanese Development Agency
KBO	Kagera River Basin Organization
LRTAP	Long-Range Transport of Atmospheric Pollutants
MAB	Man and the Biosphere (UNESCO)
NAFTA	North American Free Trade Association
NASA	National Aeronautics and Space Administration (US)
NATO	North Atlantic Treaty Organization
NCAR	National Center for Atmospheric Research
NGOs	Non-Governmental Organizations
NOAA	National Oceanic and Atmospheric Administration

NORAD	North American Aerospace Defense Command
OAS	Organization of American States
OAU	Organization of African Unity
OECD	Organization for Economic Cooperation and Development
OPEC	Organization of Petroleum Exporting Countries
ORSTOM	Office Recherche Scientifique Technique Outre-mer
PAHO	Pan-American Health Organization
PTA	Preferential Trade Area
R&D	Research and Development
SAARC	South Asian Association for Regional Cooperation
SADCC	Southern African Development Coordination Conference
SCOR	Scientific Committee on Ocean Research
SIDA	Swedish International Development Authority
SOC	IOC Regional Committee for the Southern Ocean
SPC	South Pacific Commission
SPF	South Pacific Forum
SPREP	South Pacific Regional Environment Program
START	System for Analysis Research and Training
TAC	Total Allowable Catch
TAC	Treaty for Amazon Cooperation
TOGA	Tropical Oceans Global Atmosphere
UN	United Nations
UNCED	United Nations Conference on Environment and Development
UNCLOS	United Nations Conference on the Law of the Sea
UNDP	United Nations Development Programme
UNECE	United Nations Economic Commission for Europe
UNEP	United Nations Environment Programme
UNESCO	United Nations Educational, Scientific and Cultural Organization
USAID	US Agency for International Development
USIS	US Information Service
USSR	Union of Soviet Socialist Republics (former)
VITUKI	Water Resources Research Centre (Hungary)
VNIRO	Russian Federal Research Institute of Fisheries and Oceanography
WCED	World Commission on Environment and Development
WCIRP	World Climate Impact and Response Strategies Program
WESTPAC	IOC Sub-Commission for the Western Pacific
WMO	World Meteorological Organization
WOCE	World Ocean Circulation Experiment

INDEX

AAAS, 64
acid rain, 27, 68, 141
acknowledgments, ii
ACMAD, 70
ACP, 159
adapatation strategies, 12, 70, 73
Addis Ababa, 146
Africa, 70, 100, 101, 142
  East, 118, 146
  north, 146
  south, 145, 146
  West, 118, 146
  western, 10, 64, 116
Agenda 21, 22, 75, 78, 97, 113
  Section IV, 189-193
agenda of meeting, 169-172
AGRHYMET, 71
agriculture, 35, 36, 72, 76, 93, 143, 160
AID, 140
Alagoas, 139
Altithermal, 59
Amazon, 16, 26, 35, 64, 81, 87, 136, 141
  Cooperation, Council on, 52
  Treaty for, 52, 63
  rainforests, 141
Amudarya River, 92
analogies, 25, 135
  historical, 99
Andes, 138
Anglo-Icelandic Cod Wars, 60
Angola, 145
animal wastes, 143
Antarctic, 26, 136
Antarctica, 75, 117
  Treaty, 63, 153
AOSIS, 30, 48, 151, 160, 161, 165
aquaculture, 128
aquifers, freshwater, 90
Aral Sea, 16, 192
Arctic Commission, 46
arctic summers, warm, 59
Arctic, 26, 117, 136
Argentina, 49
Armed Forces, US, 152
ASEAN, 28, 63, 71, 159, 150
Asia, 33, 100
  Central, 92
  South, 99, 150, 152
Asian Development Bank, 71, 72, 150
Asian Disaster Preparedness Center, 72
Atatürk Dam, 107
Atlantic Ocean, 109
  North, 113, 117
atmosphere, 59, 80
atmospheric pollution, 161
Australia, 123
automobiles, 39

bagasse, 143
Baltic Sea, 46, 78, 81
Bangladesh, 41, 51, 71, 150
barrier islands, 90
Bay of Bengal cyclone, 152
beach erosion, 90
Belgrade Convention, 94
Berlin, Treaty of, 101
Bhutan, 150
bibliography, 196
biodiversity, 81
biographical statements, 173-188
biomass as alternative energy source, 69
biomass burning, 68
biosphere, 70, 136
Bolivia, 49, 51
Botswana, 145
Boundary Waters Treaty of 1909, 50
Brandt Commission, 61
Brazil, 49, 51, 63, 138, 139, 140
  northeast, 138
British empire, 142, 150
Brundtland Commission, 61, 64
  Report, 61
Buenos Aires, 49
Burundi, 146

Cameroon, 51
Canada, 50, 64, 70, 71, 73, 92
canals, 89
Caribbean, 78, 116, 118, 122, 138, 140
  Action Plan, 118
CARICOM, 159, 165
CARIPOL, 118, 119
case studies, uses of, 46, 91, 92
Caspian Sea, 59
CEHI, 165
Central America, 138
Central Indian Ocean, 122
CFCs, 39, 66, 75, 154
Chaco Region, 139
Chad, 51
charcoal, 144
Chernobyl, 81
Chile, 137
China, 82, 100, 101, 123
CIDA, 140
CIECC, 138
CIES, 138
CILSS, 45
CIMH, 165
CINCWIO, 122
circumpolar boreal forests, 10, 64
citrus groves, 103
Climate Change Convention, 162
Climate Convention, and US government, 39
cloud feedbacks, 11
CMO, 165
CO_2, 11, 12, 17, 26, 33, 35, 44, 60, 66, 67, 69, 72, 76, 88, 136, 141, 162, 165
coal, 39, 76
coastal zones, 115
Cod Wars, Anglo-Icelandic, 60
Cold War, 2, 9, 19, 46, 57, 83, 102
Colombia, 51
Colorado River, i, 92

Columbia River, 92
COMECON, 63
Comision Permanente del Pacifico Sur, 113, 118
Commission for the Rhine, 100
Commission of the European Community, 154
common security, 84
Congo, 101
Congress of Vienna, 100
conservation, water, 105
consumption and population, 44
Convention for Protection of Global Atmosphere, 40
  on Long-Range Transboundary Air Pollution, 147
  on Nature Protection and Wildlife Preservation in the Western Hemisphere, 137
Cooperative Study of Kuroshio and Adjacent Regions, 116
Council on Amazon Cooperation, 52
CPPS, 113, 118
Crimean War, 100
cryosphere, 70
CSCE, 154
cyclone warning systems, 47
cyclones, 59, 152, 160

Dakotas, 50
dams, 37, 41, 50, 89, 95
Danube River, i, 18, 51, 77, 93-95
  Commission, 45, 63, 94, 96, 99, 100
data collection, 12, 20, 47, 70, 73, 95, 104, 105, 108, 118, 125
deforestation, 16, 76, 89
  tropical, 58, 68, 81, 87
Denmark, 81
depletion of stratospheric ozone, 141, 139, 148
desertification, 58
  hazards studies, 139
deserts, 103
dikes, 89
Douro River, 21, 109
DRDE, 139
droughts, 41, 69, 90, 92, 147, 160, 162
  African, 45
Dublin International Conference on Water and Environment, 97
Dublin, 53
DWFNs, 43

Earth Summit (*see also* UNCED), 63, 77
East Africa, 35, 118, 143, 144, 146
East African
  Common Services Organization, 142
  Community, 27, 142-145, 148
  Development Bank, 144
  High Commission, 142
East Asian Sea, 118
Eastern Europe, ii, 83
EC, 28, 42, 47, 51, 77, 111, 149, 157-159, 162
ECAFE, 101
ecology, 73, 94
Economic Commission
  for Africa, 70
  for Europe, 48
ecosystems, river basin, 35
Ecuador, 51

EEC, 63
EEZs, 23, 42, 43, 122, 126, 128-130
Egypt, 51, 51, 146
El Niño, 138
ENSO events, 36, 50, 70
environment, treaty on, 51
environmental
  refugees, 15
  security, 15, 80-87
epistemic communities, 121, 156
erosion, 93
  beach, 90
ESCAP, 147
ESIG, 2
estuarine resources, 35
ETAP, 101
Ethiopia, 51, 53, 146
Euphrates, i, 98, 107
Europe, 78, 95, 99, 121
  Eastern, ii, 83
  Southern, 111
European Commissions, 101
evapotranspiration, 90
extreme meteorological events, 17, 34, 59, 62, 69, 73, 92, 94, 160

famine, 82
FAO Fisheries Council and Commissions, 117
FAO, 98, 101, 125, 162, 169
FCCC, 30, 36, 38, 39, 40, 49, 58, 69, 160
FINNIDA, 140
fish
  habitats, 90
  populations, 38
  resources in Japan, 123
fisheries, 42
  near-shore, 40
Fisheries Council and Commissions, FAO, 117
Flood Action Plan (Bangladesh), 41
  warnings, 71
floodplain vegetation, 94
floods, 41, 69, 89, 90, 92, 94, 162
Florida, 60
food security, 80
forecasting by analogy, 59, 99, 102
  types of, 19
forecasting systems, 95
forests, 10, 35, 36, 42, 76, 81, 143, 160, 161
  circumpolar boreal, 10
fossil fuels, 35, 39, 68, 72, 76, 81, 138, 143, 144
Framework Convention on Climate Change, 30, 36, 38, 39, 40, 49, 58, 69, 160
France, 123
freezes in Florida, 60
freshwater aquifers, 90
frost, 41
fuel wood, 82, 144

Ganges-Brahmaputra, 18, 96
  basin, 51
Garrison Diversion Project, 50
GATT, 44, 63, 77, 97, 161
GCMs, 17, 33, 34, 40, 58, 59, 92
  and policymaking, 34
GEC (global environmental change), 15-16, 81, 87

general circulation models, ii
Geneva, 147
GESAMP, 114
GEWEX, 70
GHGs, 11, 12, 33, 39, 44, 47, 49, 66-69, 72, 76, 127
Global Environmental Facility, 70
global thermonuclear war, 84
Golan Heights, 107
Great Lakes, 35, 50, 59, 63, 71, 77-78
greenhouse effect, 66
greenhouse gases, 11, 12, 33, 39, 44, 47, 49, 66-69, 72, 76, 127
Group of 77, 150, 151, 160
Guadiana River, 21, 109
Guyana, 51
Guyanese Shelf, 35

hazardous wastes, 137
Helsinki
  Commission, 118
  Protocol, 48, 147
historical analogs, ii
Human Dimensions of Global Environmental Change Program, 48
Hungary, 41
hurricanes, 41
hydraulic works, 111
hydroelectricity, 111, 143, 144
hydrologic cycle, 34, 67, 162
hydrometeorological observations, 94

Iberian Peninsula, 21, 51, 109, 111
Ice Age, 57
Iceland, 60
ICES, 22, 113, 118, 156
ICSPRO, 114
ICSU, 36, 64
IDB, 140
IDNDR, 72
IGBP, 48, 70
IGOs, 153-154, 156-157
IIASA, 64
IJC, 50, 63, 72, 78
IMF, 77
impact assessments, 12, 70, 73
INC, 40, 48
India, 51, 53, 71, 100, 149, 150, 151
Indian Ocean, 117
  Expedition, 114, 116
  Marine Affairs Cooperation, 118
  Central, 122
  southeast, 147
  Western, 122, 124
Indonesia, 71
Indus River, i
  basin, 98
Indus River Commission, 101, 149
institutional lag, 13, 76, 78, 80
Inter-American Institute, 70
  for Global Change Research, 138
Inter-American Program of Action for Environmental Protection, 137
Intergovernmental Coordinating Committee, 49
International Commission for the Congo, 101
International Conference on Water and Environment, 19, 53, 102
International Council for the Scientific Exploration of the Mediterranean, 118
International Indian Ocean Expedition, 116
International Institute for Agriculture, 101
International Law Commission, 19, 102
International Maritime Organization, 118
International Meteorological Organization, 101, 113
International Tropical Timber Organization, 63
International Whaling Commission, 77, 125, 153, 157, 158
IOC, 22, 36, 113-118, 120, 122, 160, 162
IOCARIBE, 116, 118, 122
IOCEA, 116
IOCINCWIO, 116, 117
IOCINDIO, 116, 117
IPCC, 12, 33, 34, 40, 47, 48, 64, 68, 127, 161, 162, 165
  Working Group I, 39
Iraq, 57, 80, 107
irrigation systems, 105
Israel, 107
ISSC, 48
IWC, 77, 125, 153, 157, 158

Japan, 123
JICA, 140
Jordan River basin, 98, 107

Kagera River Basin Organization, 146
Kashmire, 151
Kathmandu, 151
KBO, 146
Keflavik, 60
Kenya, 123, 142, 144, 135
Kuroshio, 116
Kuwait, 57, 80

La Plata, 18, 49, 96, 136
  river basin, 138
Lake Baikal, 16, 81, 87
Lake Balaton, 41
Lake Chad Basin Commission, 51
Lake Kyoga, 146
Lake Victoria, 51, 142, 146
  basin, 34, 51
Lake Winnipeg, 50
land use, 160
Latin America, 41, 138, 140
Law of Non-Navigable Uses of International Watercourses, 97
Law of the Sea, 43
  Conference, Third, 126 (*see also* UNCLOS)
  Convention, 114
League of Nations, 101
Lesotho, 145
list of participants, 195
Litani River, 107
LRTAP, 48, 147
Lusaka, 145, 146

Mackenzie River, 70
Malawi, 145
Malaysia, 71
Maldives, 150, 151
Mamoré, 139
mangroves, 90
Mannheim, Treaty of, 100

Mato Grosso, 138
measurement standardization, 39
Mediterranean, 118
  basin, 33
  Blue Plan, 78
  Sea, 109
Mekong
  basin, 98
  Committee, 98
  River Commission, 101, 149
methane, 66, 72, 76
Mexico, 73, 92, 122, 138
Middle East, 20, 40, 82, 104, 106, 108
migration, 82
Migratory Birds Treaty in North America and Europe, 63
Mississippi, 136
  River, 70
  watershed, 50
Missouri basin, 50
Mobutu Sese Seko, 146
models, climate, 11, 67, 68
  ocean-atmosphere, 33
  regional, 34
Mombasa, 142
monsoons, 33
Montreal Protocol, 75, 137
mountains, 97
Mozambique, 145
multilateral aid, 120

NAFTA, 63, 77
Namibia, 145
National Environmental Policy Act of 1969, 42
NATO, 1, 46, 60, 63, 64, 154, 159
  ARW, 62
Nature Conservancy, 140
navigation, 94
NCAR, 1, 2
Nepal, 51, 53, 150, 151
New Caledonia, 123
New Delhi, 151
NGOs, 13, 14, 61, 78, 79, 80
  Canadian, 50
  role of, 53-54
  US, 50
Niger, 18, 51, 96, 101
  Basin Commission, 51
Nigeria, 51
Nile, 18, 51, 96, 107, 142, 144
nitrous oxides, 48, 66, 147
NOAA, 139
Non-Aligned Group, 150
North Africa, 146
North America, 73, 99
North Atlantic, 113, 117
  Alliance, 60
North Pacific Fur Seal Commission, 125
North Sea, 78, 8
  pollution of, 121
Northeast
  Atlantic, 36
  Pacific, 36
nuclear testing, French, 123
nuclear war, 86

OAS, 26, 28, 52, 72, 136, 137, 139-141, 149, 159
  Charter, 136
  General Secretariat, 138
OAU, 72, 146
ocean
  circulation, 24
  -atmosphere models, 33
oceans, 10, 24, 59, 70, 120
  health of, 65
OECD, 47, 63, 72, 151, 156, 157, 159, 161
oil, 39
  prices, increase in, 144
  resources in Japan, 123
  wells, 80
OPEC, 63
Orontes River, 107
Ottoman Empire, 63
overcapacity, 128
overconsumption, 44
overfishing, 24, 42, 43, 138-130
Owen Falls Dam, 144
ozone depletion, 39, 57, 58, 68, 75, 137, 141, 148, 156

Pacific Ocean, southeast, 138
Pacific
  South, 147, 162
  Southeast, 118
PAHO, 72
Pakistan, 71, 149-151
paleoclimatic reconstructions, ii, 59
Palmer Commission, 84
Pan-American Union, 101
Paraguay, 49, 139
  River, 50
Parana River, 50
ParapetíRiver basin, 139
Paris, 122
  Treaty of, 100
pelagic fish, 36
Persian Gulf War, 57
Peru, 51
pesticides, 143
petroleum, 76
Philippines, 71
photovoltaic energy, 69
Pohai Gulf, 123
pollutants, 41, 117
pollution
  control, 98, 108
  North Sea, 121
  atmospheric, 161
  land-based, 40
  transboundary water, 71
populations, 73, 76
  and consumption, 44
Portugal, 21, 92, 109, 111, 112
power generation, 93
Preferential Trade Area, 146

rainfall, 50, 162, 165
rainforests, 141
rainforests, tropical, 59
Red River, 50
reforestation, 35
Regional Committee for the Southern Ocean, 117
regional development banks, 46
Regional Meteorological Centre, 70
Regional Seas Programme, 70, 73, 117

reservoirs, 89, 96
resource scarcity, 24, 43, 128
Rhine, 18, 77, 95
  Commission, 63, 99, 100
Rio de Janeiro, 63
river basins, 35, 139
  organizations, 46
Roman Empire, 63
Romania, 41
Rome Declaration, 15
runoff, 89, 93
Rwanda, 146

SAARC, 28, 149-159
SADCC, 45, 145, 147
Sahelian droughts, 45
Salt Lake, 59
Santiago, 137
  Declaration, 113
Saõ Francisco River basin, 139
scarcity of resources, 24, 43, 128
SCOR, 36
sea level rise, 42, 71, 73, 89, 90, 115, 165
seas, 10, 80, 128
  health of, 65
security
  definition of, 15
  economic, 80
  food, 80
  humanitarian, 80
Seychelles, 123
Singapore, 70
SO_2, 48
SOC, 116
Sofia Protocol, 147
soil, 108
solar energy, 69
Souris River, 50
South Africa, 145, 146
South America, 26, 41, 136, 141
South Asia, 28, 99, 150, 152
  Sea, 118
South Pacific, 36, 147, 162
Southeast
  Asia, 28
  Indian Ocean, 147
Southeast Pacific Ocean, 118, 138
Southern Ocean, 75, 117
Spain, 21, 92, 109, 111, 112
SPC, 165
SPF, 159
SPREP, 70, 165
Sri Lanka, 71, 150, 151
START Program, 47-48, 70
storm surges, 69, 90
Sudan, 146
  southern, 51
sufficiency, environmental, 83
Suriname, 51
sustainable development, 10, 38, 83, 84, 86, 113
  of global forests, 65
Swaziland, 145
Syrdarya River, 92
Syria, 107

Tagus River, 21, 109
Tamil, 151
Tanganyika, 101, 142
Tanzania, 123, 142, 144, 145, 146
technology transfer, 44, 69, 72, 124
thermal energy, 69
Third World, 86
Tigris-Euphrates, 18, 96, 107
TOGA, 70, 119
topsoil, 81
tornados, 41
total allowable catch, 158
tourism, 138
tragedy of the commons, 125
transboundary water pollution, 71
Treaty
  for Amazon Cooperation, 51, 139
  of Berlin, 101
  of Mannheim, 100
  of Paris, 100
  on environment, 51
  on water, 51
Tropical Cyclone
  Committee, 147
  Project, 147
tropical deforestation, 87
tuna stocks, 36
Turkey, 107
Typhoon Committee, 147

Uganda, 51, 142, 144, 146
  Railway, 142
UK, 60
UN, 45, 48, 57, 63, 70, 77, 83, 96, 101, 117, 119, 125-126, 147, 150, 160, 162
  Declaration on Human Rights, 83
  Economic Commission for Europe, 72, 147
  General Assembly, 160
  International Law Commission, 97
  Law of the Sea Convention, 114
  Security Council, 57, 159
UNCED, 14, 19, 22, 44, 49, 63, 64, 69, 75, 77, 78, 80, 97, 102, 113, 140, 151, 152, 158, 161
UNCLOS, 113, 119, 156
UNDP, 51, 119, 140, 160
UNEP, 64, 68, 70, 73, 82, 113, 114, 118, 140, 160, 162
  Caribbean Action Plan of, 118
  Regional Seas Programme of, 70, 73, 117
UNESCO, 122, 136, 160
urban water systems, 105
urbanization, 89
Uruguay, 49, 97
US, 44, 50, 57, 64, 70, 71, 73, 75, 76, 86, 92, 123, 138
  government and Climate Convention, 39
  western, 34
USAID, 140
USIS, 1
USSR, ii, 46, 57, 83, 86, 123
utilities, 39

vegetation, floodplain, 94
Venezuela, 51, 122
Vienna Convention, 75
Vietnam, 71
  War, 101
volcanic eruptions, 66

war, thermonuclear, global, 84
warning systems, 95

Warsaw Pact, 63
waste-water treatment, 95
wastes
  animal, 143
  hazardous, 137
water, 82
  allocation, 40
  conservation, 20, 105, 108
  discharge, 108
  drinking, 93
  pollution, transboundary, 71
  resources, 38
  systems, 105
  treaty on, 51
  use, more efficient, 20
  vapor, 11
  wars, 17, 92, 98
watersheds, Portuguese-Spanish, 109
WCED, 61
WCRP, 70
West Africa, 118, 146
Western Hemisphere Convention, 137
Western
  Indian Ocean, 122
  Pacific Ocean, 123
WESTPAC, 116, 123
wetlands, 90, 138
whaling olympics, 158
WHO, 72
wind, 69
winners and losers, 38, 44, 58
WMO, 36, 47, 51, 68, 70, 101, 113, 147, 160, 162
Working Group I, 33
World Bank, 63, 101, 107, 117
World Climate Program, 48

Yangtse River, 101

Zaire, 146
Zambia, 145, 146
Zimbabwe, 145

Printing: Druckhaus Beltz, Hemsbach
Binding: Buchbinderei Schäffer, Grünstadt

Springer-Verlag and the Environment

We at Springer-Verlag firmly believe that an international science publisher has a special obligation to the environment, and our corporate policies consistently reflect this conviction.

We also expect our business partners – paper mills, printers, packaging manufacturers, etc. – to commit themselves to using environmentally friendly materials and production processes.

The paper in this book is made from low- or no-chlorine pulp and is acid free, in conformance with international standards for paper permanency.